Sustainable Infrastructure impacts our well-being and day-to-day lives. The infrastructures we are building today will shape our lives tomorrow. The complex and diverse nature of the impacts due to weather extremes on transportation and civil infrastructures can be seen in our roadways, bridges, and buildings. Extreme summer temperatures, droughts, flash floods, and rising numbers of freeze-thaw cycles pose challenges for civil infrastructure and can endanger public safety. We constantly hear how civil infrastructures need constant attention, preservation, and upgrading. Such improvements and developments would obviously benefit from our desired book series that provide sustainable engineering materials and designs. The economic impact is huge and much research has been conducted worldwide. The future holds many opportunities, not only for researchers in a given country, but also for the worldwide field engineers who apply and implement these technologies. We believe that no approach can succeed if it does not unite the efforts of various engineering disciplines from all over the world under one umbrella to offer a beacon of modern solutions to the global infrastructure. Experts from the various engineering disciplines around the globe will participate in this series, including: Geotechnical, Geological, Geoscience, Petroleum, Structural, Transportation, Bridge, Infrastructure, Energy, Architectural, Chemical and Materials, and other related Engineering disciplines.

More information about this series at http://www.springer.com/series/15140

Sherif Badawy · Dar-Hao Chen
Editors

Recent Developments in Pavement Engineering

Proceedings of the 3rd GeoMEast
International Congress and Exhibition, Egypt
2019 on Sustainable Civil Infrastructures –
The Official International Congress
of the Soil-Structure Interaction Group
in Egypt (SSIGE)

 Springer

Editors
Sherif Badawy
Mansoura University
Mansoura, Egypt

Dar-Hao Chen
Texas A&M University
College Station, TX, USA

ISSN 2366-3405   ISSN 2366-3413   (electronic)
Sustainable Civil Infrastructures
ISBN 978-3-030-34195-4   ISBN 978-3-030-34196-1   (eBook)
https://doi.org/10.1007/978-3-030-34196-1

This Springer imprint is published by the registered company Springer Nature Switzerland AG
The registered company address is: Gewerbestrasse 11, 6330 Cham, Switzerland

Contents

About the Editors

Sherif El-Badawy Ph.D, Associate Professor, Mansoura University, Public Works Engineering Department, Faculty of Engineering, Mansoura University, Mansoura, Egypt, PH: +201000183519, E-mail: sbadawy@mans.edu.eg

Sherif El-Badawy is Associate Professor at Mansoura University, Egypt. He serves as the Director of Highway and Airport Engineering Laboratory, Director of the Center of Scientific, Experimental, and Technical Services, Mansoura University. He received a BSc degree with honor in Civil Engineering and MSc from Mansoura University, Egypt. Dr. El-Badawy pursued his PhD Degree at Arizona State University (ASU), USA, in 2006. He worked as a Postdoctoral Research Associate at ASU from June 2006 until July 2007 and as a Research Fellow at the University of Idaho from December 2009 until January 2012. He has more than 20 years of experience in pavement structural analysis, design, and characterization. Dr. El-Badawy has participated in several state and national projects during his employment in the USA. He was one of the team members under Professor Witczak in the NCHRP 1-37A and 1-40D Projects which focused on the development of the Mechanistic-Empirical Pavement Design Guide (MEPDG). He serves as a TRB Committee Member on Flexible Pavement Design (AFD60) since April 2012. He is an elected board member of the Middle East Society of Asphalt Technologists (MESAT), a member of the International Geosynthetic Society (IGS), Transportation and Traffic Society (TTS), International

Association of Computer Science and Information Technology (IACSIT), and Chi-Epsilon Honor Society. He developed and taught many undergraduate and postgraduate courses in the USA and Egypt. His research interest focuses on pavement material characterization and modeling, Mechanistic-Empirical pavement design methods, and traffic characteristics. Dr. EL-Badawy was the Chief Editor of the ASCE GPS "Innovative Technologies for Severe Weathers and Climate Changes." Through research and graduate advising, he published more than 70 technical publications and reports.

Dar Hao Chen Ph.D. PE
13208 Humphrey Dr. Austin TX 78729
+1(512)705-6263
d-chen@tti.tamu.edu
Darhao2008@gmail.com

Activities

- Editor In Chief, Innovative Infrastructure Solutions, Springer
- Associate Editor, Journal of Performance of Constructed Facilities, ASCE
- Associate Editor, International Journal of Geomechanics, ASCE
- Member of Editorial Board, Journal of Testing and Evaluation, ASTM
- Board Director, ILISO Consulting Zambia Limited, Zambia
- Panel Member for National Academy, National Cooperative Highway Research Program (NCHRP), Transportation Research Board, Washington D.C.
- President, GeoChina Civil Infrastructure Association, USA
- Committee Member AFD70, Pavement Rehabilitation. Transportation Research Board, Washington D.C.
- Committee Member AFH30, Emerging Technologies for Design and Construction. Transportation Research Board, Washington D.C.
- Board Member, SSIGE, Egypt

- Committee Member, AFS90 for Chemical and Mechanical Stabilization, Transportation Research Board, Washington D.C.
- Steering Committee Member on ASCE Geo-Institute's Pavements Committee
- Steering Committee Member on ASCE Geo-Institute's Geophysical Engineering Committee
- Committee Member AFS50 for Modeling Techniques in Geomechanics, Transportation Research Board, Washington D.C.
- Committee Member AFS20 for Soils and Rock Instrumentation. Transportation Research Board, Washington D.C.
- Committee Member for Warm Mix Asphalt Standard Specification in China
- Co-Chairs and Secretary Generals of five International Conferences—2009 and 2011 GeoHunan, 2014 GeoHubei, 2016 and 2018 GeoChina International Conferences
- Adjunct Professor (1) University of Oklahoma, USA (2) Shandong University, China (3) Zhejiang University, China (4) Beijing Jiaotong University, China (5) Changsha University of Science and Technology, China (6) National United University, Taiwan

Education

PhD Civil Engineering | University of Oklahoma, 1994

Experiences

Senior Research Associate | TTI, Texas A&M University

03/2018 – Present
Road Materials and Pavement Engineering. Forensic Studies of Roadway Premature Failures. Pavements (Design, Construction, QC/QA, Performance Testing, Management, and Rehabilitation). Optimum rehabilitation strategy for roadway pavements.

**APT System Manager and Pavement Engineer |
Texas Dept of Transportation**

03/1995 – 02/2018
Developed and conducted optimum rehabilitation
and maintenance strategies for over 600 projects.
They included both flexible and rigid pavements
through comprehensive material characterizations
using the state-of-the-art non-destructive testing
(NDT) techniques. Performed more than 500 forensic
studies and other special projects using NDT as well as
advanced laboratory testing to analyze and characterize
pavement sections. Directed, monitored, and super-
vised Texas DOT's research projects and IAC contracts
as a Project Director, Project Advisor, and Project
Manager with funding exceeds $200 million dollars.

Research Experiences

1. 5-6271-05 TxDOT Research Project: Implementing
   Full Depth Reclamation (FDR) to Repair Roads
   Damaged in the Energy Sector. 2015–2017-
   $1,043,000
2. 0-6856, TxDOT Research Project: Sustainable
   Perpetual Asphalt Pavements and Comparative
   Analysis of Lifecycle Cost to Traditional 20-Year
   Pavement Design. 2015–2018-$572,426.
3. 0-6839 TxDOT Research Project: Designing
   Pavements to Support the Heavy Loads in the
   Energy Development Areas. 2015–2018-$527,647.
4. 0-6854 TxDOT Research Project: Engineering the
   Properties of Asphalt Mixtures Using Carbon
   Nanotubes, 2015–2016-$265,438.
5. 5-5598-05 TxDOT Research Project: Continued
   Implementation of High Performance Thin
   Overlays in Texas Districts. 2014–2016-$360,000
6. 0-5836 TxDOT Research Project: Performance of
   Permeable Friction Course (PFC) Pavements Over
   Time. 2008–2012-$663,214
7. 0-6005 TxDOT Research Project: Developing a
   Testing Device for Total Pavements Acceptance.
   2008–2011-$1,411,570
8. 0-6092 TxDOT Research Project: Performance
   Evaluation and Mix Design for High RAP
   Mixtures. 2008–2011-$425,975

9. 0-6611 TxDOT Research Project: Improvements of Partial and Full-Depth Repair Practices for CRCP Distresses. 2010–2012-$269,255

10. 0-6613 TxDOT Research Project: Evaluate Binder and Mixture Aging for Warm Mix Asphalts. 2010–2013-$525,511

11. 0-6190 TxDOT Research Project: Use of Dowel Bars at Longitudinal Construction Joints. 2009–2010-$250,000

12. 0-5270 TxDOT Research Project: A Logical Guideline for Superheavy Load Review Policy. 2008–2009-$245,067

13. 0-5444 TxDOT Research Project: Rehabilitation Procedures for Longitudinal Cracks and Joints Separation in Concrete Pavement. 2005–2009-$550,000

14. 0-4517 TxDOT Research Project: Develop Statewide Recommendations for Application of PCC Joint Reflective Cracking Rehabilitation Strategies Considering Lufkin District Experience. 2003–2005-$240,550.

15. SPR-2(205) Model Calibrations with Local Accelerated Pavement Test Data and Implementation for Focused Solutions to NAFTA Problems. 2001-2007-$960,000.

16. TPF-5(004) Long-Term Pavement Performance (LTPP) Specific Pavement Study (SPS) Traffic Data Collection. 2004-2009-$2,925,390.

17. SPR-2(208) Pavement Subgrade Performance Study. 2001–2006-$2.59 million.

18. 0-4822 Monitor Field Performance of Full Depth Asphalt Pavements to Validate Design Procedures. 2004–2009-$906,370.

19. 0-5821 Develop Guidelines for Routine Maintenance of Concrete Pavement. 2006–2008-$145,910.

20. 0-6005 Developing a Testing Device for Total Pavements Acceptance. 2009–2010-$1,411,570.

21. 0-5836 Performance of Permeable Friction Course (PFC) Pavements Over Time. 2009–2011-$663,214.

22. 0-6192 Performance Evaluation and Mix Design for High RAP Mixtures. 2009–2011-$387,975

23. 0-5549 Horizontal Cracking in Concrete Pavements. 2006–2008-$446,023.
24. 0-5513 Development of a Flexible Pavements Database. 2005–2008-$450,000.
25. 0-5472 A Data Base for Successful Pavement Sections in Texas - Including Experimental and Non-Experimental Pavements. 2005–2007-$280,000.
26. 464PVIA005 Data Analysis and Instrumentation Support. 2005–2007-$180,000.
27. 463PVIA003-NDT Data Analysis for Special Projects. 2005–2007-$60,000.
28. 465PVIA005-Rolling Dynamic Deflectometer (RDD) and Stationary Dynamic Deflectometer (SDD) Support. 2005–2009-$450,000.
29. 465PVIA007-GIS support for Load Zone Analyses in Texas. 2005–2009-$250,000.
30. 462XXA1005 Pavement Performance Evaluation of MLS Test Sites. 2001–2003-$130,000.
31. 462XXIA017 Superheavy Load and Load Zone Analysis Support. 2001–2003-$100,000.
32. 463PVIA001 Laboratory & NDT support for MLS testing. 2001–2003-$120,000.
33. 463PVIA002 GIS Support for Load Zone Analyses in Texas. 2001–2003-$100,000.
34. 463PVIA003 Lab & NDT support on special projects. 2003–2009-$250,000.
35. 464PVIA-NDT & Laboratory Support for Ft. Worth SH114 and Other Special District Projects. 2003–2009-$350,000.

Patents

First Author of Five (5) patents in China

Publications-Editor

1. GeoChina 2016 "Sustainable Civil Infrastructure—Innovative Technologies and Materials". Edited by Dar-Hao Chen, Hany Farouk Shehata, and Zhanyong Yao. ASCE Geotechnical Special Publications (GSP) 257–268. CD-ROM. 2237 pp. ISBN 9780784480113. 2016.

2. GeoHubei 2014 "Sustainable Civil Infrastructure—Innovative Technologies and Materials". Edited by Dar-Hao Chen and Shu-Rong Yang. ASCE Geotechnical Special Publications (GSP) 244–255. CD-ROM. 1543 pp. ISBN 9780784413593. 2014.

3. ASCE International Journal of Geomechanics "Modeling in Geotechnical Engineering for Design and Materials" edited by Rafiqul A. Tarefder, Dar-Hao Chen, and Yu-Ning Ge. Volume 10, Issue 6, pp. 213–249. ISSN 1943-5622 (online). 1532-3641 (print). Nov/December 2010.

4. ASCE Geotechnical Special Publication 218 "Emerging Technologies for Material, Design, Rehabilitation, and Inspection of Roadway Pavements edited by Dar-Hao Chen, Jia-Ruey Chang, Musharraf Zaman, Chaoyang Zhao, and Zhanyong Yao. ISBN 978-0-7844-7629-1. 2011.

5. ASCE International Journal of Geomechanics "Advanced Modeling and Numerical Simulations in Geomechanics" edited by Yu-Ning Ge, Dar-Hao Chen, and Rafiqul A. Tarefder. Volume 11, Issue 3. May/June 2011. ISSN 1943-5622 (online). 1532-3641 (print).

6. GeoHunan 2011 "Emerging Technologies for Design, Construction, Rehabilitation, and Inspection of Transportation Infrastructures". Edited by Dar-Hao Chen and Hongyuan Fu. ASCE Geotechnical Special Publications (GSP) 212–223. CD-ROM. 2945 pp. ISBN 9780784411742. 2011.

7. ASCE Geotechnical Special Publication 193 "Material, Design, Construction, Maintenance, and Testing of Pavement" edited by Chen, Dar-Hao, Estakhri, C., Zha, S., and Zeng, S. ISBN 978-0-7844-1045-5. 2009.

Consultation Services

- Corsair Consulting LLC, Austin TX
- Beyond Engineering and Testing, Austin TX
- Shandong Transportation Research Institute, China
- Chongqing Pengfang Pavement Engineering Inc., China
- Corasfaltos, Colombia
- Shanghai RMI Pavement Engineering Inc., China

- Taiwan Taoyuan International Airport–Civil Aviation Corporation, Taiwan
- Lily Corporation, USA
- Gifford-Hill & Company, USA
- Qinghai Transportation Research Institute, China
- Transportation Research Institute of Jiangsu Province, China
- Highway Planning Survey and Design Institute Sichuan Communications Department, China
- China Construction Machinery Association, Shandong, China
- Guangzhou Transportation Research Institute, China
- In cooperation with Changsha University of Science and Technology and NCHRP on translating NCHRP Report 691 "Mix Design Practices for Warm Mix Asphalt" to Chinese
- In cooperation with Cangzhou Engineering Company (China) and Meeker Equipment (USA), successfully installed 10 warm mix producing equipment through foaming technology in Hebei, Jiangsu, Shandong, Shanxi, and Liaoling provinces
- In cooperation with Cangzhou Engineering Company (China) and Hebei Science and Technology Bureau, successfully built experiment roads by using warm mix foaming technology
- Invited speaker to make presentations related to warm mix technologies at the Chinese Ministry of Transportation, Highway Associations, Transportation Research Institutes and numerous Universities throughout China

Journal Papers

Author of more than 150 papers in the following 16 technical journals

1. Construction and Building Materials
2. Journal of Performance of Constructed Facilities, ASCE
3. NDT & E International Journal
4. Journal of Testing and Evaluation, ASTM
5. Arabian Journal of Geosciences

6. Canadian Journal of Civil Engineering
7. International Journal of Road Materials and Pavement Design
8. International Journal of Pavement Engineering
9. KSCE Journal of Civil Engineering
10. Journal of Geotechnical and Geoenvironmental Engineering, ASCE
11. Transportation Research Record: Journal of the Transportation Research Board
12. Journal of Materials in Civil Engineering, ASCE
13. Journal of Transportation Engineering, ASCE
14. Geotechnical Testing Journal, ASTM
15. Journal of Infrastructure Systems, ASCE
16. International Journal of Computational Intelligence Research

SCI Indexed Journals

1. Kai Yao, Na Li, and Dar Hao Chen (2019) "Generalized Hyperbolic Formula Capturing Curing Period Effect on Strength and Stiffness of Cemented Clay". Construction and Building Materials, Volume 199, January 2019, Pages 63–71.
2. Kai Yao, Huawen Xiao, Dar-Hao Chen, Yong Liu (2019) "A Direct Assessment for the Stiffness Development of Artificially Cemented Clay" Géotechnique. ISSN 0016-8505 | E-ISSN 1751-7656.
3. Dar Hao Chen, Tom Scullion, and Boo Hyun Nam (2016) "Characterization of Structural Conditions for Pavement Rehabilitations". Construction and Building Materials, Volume 121, 15 September 2016, Pages 664–675.
4. Dar Hao Chen, Carlos Lam, and Jinyuan Liu "Special Issue on Sustainable Civil Infrastructures: Innovative Technologies and Materials" Journal of Performance of Constructed Facilities, ASCE. Vol 30, Issue 1, pp 1–3. Feb-2016.
5. Dar Hao Chen, Pangil Choi, Kuan Yu Chen, Moon Won (2016) "Slot stitching for longitudinal joint separation repairs". Construction and Building Materials, Volume 115, 15 July 2016, Pages 153–162.

6. Dar Hao Chena, Moon Won, Xianhua Chen, and Wujun Zhou (2016) "Design improvements to enhance the performance of thin and ultra-thin concrete overlays in Texas". Construction and Building Materials. Volume 116, 30 July 2016, Pages 1–14.

7. Chen, Dar-Hao, and Hong, F. (2015) "Long Term Performance of Diamond Grinding" Journal of Performance of Constructed Facilities, ASCE. Vol. 21, No. 1, pp 06014006-1–06014006-7. Jan/Feb, 2015.

8. Hong, F. and Chen, Dar-Hao (2015) "Evaluation of Asphalt Overlay Permanent Deformation Based on Ground Penetrating Radar Technology" Journal of Testing and Evaluation, ASTM. Volume 43, Issue 4. July 2015.

9. Chen, Dar-Hao, and Won, M. (2015) "Field Performance with State-Of-The Art Patching Materials" Construction and Building Materials. Volume 93. September 2015, Pages 393–403.

10. Chen, Dar-Hao, Tang, C., Xiao, HB, Ying, P. (2015) "Utilizing electromagnetic spectrum for subsurface void detection—case studies. Arabian Journal of Geosciences. Volume 8, Issue 9, pp 7705–7717. September 2015.

11. Chen, Dar-Hao, and Scullion, T. (2015) "Very Thin Overlays in Texas" Construction and Building Materials. Volume 95. October 2015, Pages 108–116.

12. Chen, Dar-Hao, and Won, M. (2015) "CAM and SMA Mixtures to Delay Reflective Cracking on PCC Pavements" Construction and Building Materials. Volume 96. October 2015, Pages 226–237.

13. Chen, Dar-Hao, and Yi, W. (2015) "Performance of Settled Bridge-Approach. Slabs with Polyurethane-Foam Injection" Journal of Testing and Evaluation, ASTM. Vol. 43. No. 6. November 2015.

14. Chen, Dar-Hao, Zhou, W., Yi, W., and Won M. (2014) "Full-depth Concrete Pavement Repair with Steel Reinforcements" Construction and Building Materials. Volume 51, 31 January 2014, Pages 344–351.

15. Chen, Dar-Hao, and Hong, F. (2014) "Bonded Continuously Reinforced Concrete Overlay on Jointed Concrete Pavement". Canadian Journal of Civil Engineering. 41(5): 432–439, May 2014.

16. Chen, Dar-Hao (2014) "Repairs of Longitudinal Joint Separations and Their Performances" Construction and Building Materials. Volume 54, 15 March 2014, Pages 496–503.

17. Xu C., Xu C., Chen Q., Sun F., and Chen, Dar-Hao (2014) "Applications of Electromagnetic Waves for Void and Anomaly Detections". Journal of Testing and Evaluation, ASTM. Volume 42, Issue 2. March 2014.

18. Chen, Dar-Hao, Sun, R, and Yao, Z. (2013) "Impacts of Aggregate Base on Roadway Pavement Performances" Construction and Building Materials. Volume 48, November 2013, Pages 1017–1026.

19. Chen, Dar-Hao, Zhou, W., and Li, K. (2013) "Fiber Reinforced Polymer Patching Binder for Concrete Pavement Rehabilitation and Repair" Construction and Building Materials. Volume 48, November 2013, Pages 325–332.

20. Chen, Dar-Hao, Scullion, T., Hong, F., and Lee, J. (2012) "A study on the Pavement Swelling and Heaving at State Highway 6". Journal of Performance of Constructed Facilities, ASCE. Vol. 26, No. 3, pp 335–344. May/June, 2012.

21. Chen, Dar-Hao, Hong, F. and Yuan, J. (2012) "Effect of Tie Bars on the Field Performance of Full-Depth Repair on Concrete Pavement" International Journal of Road Materials and Pavement Design. Volume 13, Issue 1, pages 12–25. March 2012.

22. Hossain, Z., Zaman, M., O'Rear, EA. and Chen, Dar-Hao (2012) Effectiveness of water-bearing and anti-stripping additives in warm mix asphalt technology. International Journal of Pavement Engineering 13 (5), pages 424–432.

23. Chen, Dar-Hao, and Paulo Cruz (2012) "Performance of Transportation Infrastructure" Journal of Performance of Constructed Facilities Apr 2012, Vol. 26, No. 2, pp. 136–137.

24. Zhou, F., Sheng Hu, Chen, Dar-Hao, and Scullion, T. (2012) "RDD Data Interpretation and Its Application on Evaluating Concrete Pavements for Asphalt Overlays" Journal of Performance of Constructed Facilities, ASCE. Vol. 26, No. 5, pp 657–667. October, 2012.

25. Chen, Dar-Hao, Chang, G., and Fu, H. (2011) "Limiting Base Moduli to Prevent Premature Pavement Failure". Journal of Performance of Constructed Facilities, ASCE. Vol. 25, No. 6, pp 490–498. December, 2011.

26. Chen, Dar-Hao, Hong, F., and Zhou, F (2011) "Premature Cracking from the Cement Treated Base and Their Mitigation" Journal of Performance of Constructed Facilities, ASCE Volume 25, Issue 2. pp 113–120. April, 2011.

27. Chen, Dar-Hao, Lin, H., and Sun, R. (2011) "Field Performance Evaluations of Partial Depth Repairs" Construction and Building Materials. Volume 25, Issue 3, Pages 1369–1378. March 2011.

28. Chen, Dar-Hao, Won, M., and Hong, F. (2011) "Dowel Bar Retrofit (DBR) Performance in Texas" Construction and Building Materials. Volume 25, Issue 4, Pages 1762–1771. April 2011.

29. Chen, Dar-Hao, Hong, F., Bilyeu, J., and Yao, Z. (2011) "Improving the Performance of Full-Depth Repairs by Understanding the Failure Mechanisms. Volume 15, Number 1, 91–99. KSCE Journal of Civil Engineering. January 2011.

30. Jeong Ho Oh, Chen, Dar-Hao, Walubita, L.F., Wimsatt, A. Mitigating seal coat damage due to superheavy load moves in texas low volume roads. "Construction and Building Materials. Volume 25, Issue 8, August 2011, Pages 3236–3244.

31. Chen, Dar-Hao, and Hong, F. (2010) "Lessons Learned from RAP Sections with 17 Years of Service" Journal of Testing and Evaluation, ASTM. Vol. 38. Number 4. pp 482–493. July 2010.

32. Chen, Dar-Hao, Nam, B.H, and Yao, Z. (2010) "Utilizing Advanced Characterization Tools to Prevent Reflective Cracking" Journal of Performance of Constructed Facilities, ASCE. Volume 24, Number 4. pp. 390–398. August 2010.

33. Wang, R., Zhou, F., Chen, Dar-Hao, Zheng, G., Scullion, T., and Walubia, L. (2010) "Characterization of Rutting (Permanent Strain) Development of A-2-4 and A-4 Subgrade Soils Under HVS Loading" Journal of Performance of Constructed Facilities, ASCE. Volume 24, Number 4. pp. 382–389. August 2010.

34. Chen, Dar-Hao (2010) "Slippage Failure of a New Hot-Mix Asphalt Overlay" Journal of Performance of Constructed Facilities, ASCE. Volume 24, Number 3, pp. 258–264. June 2010.

35. Chen, Dar-Hao and Lin, H. (2010) "Effects of Base Support and Load Transfer Efficacy (LTE) on Portland Concrete Pavement Performance" Journal of Testing and Evaluation, ASTM. Vol. 38, No. 1, pp. 47–56. January 2010.

36. Chen, Dar-Hao, and Wimsatt, A. (2010) "Inspection and Condition Assessment Using Ground Penetrating Radar (GPR)" **Journal of Geotechnical and Geoenvironmental Engineering** ASCE. Vol. 136, Issue 1, pp 207–214. January 2010.

37. Hong, F., and Chen, Dar-Hao (2010) **"Long-Term Performance Evaluation of Recycled Asphalt Pavement Based on Texas LTPP SPS5 Sections"** Transportation Research Record: Journal of the Transportation Research Board. Number 2180. pp 58–66.

38. Chen, Dar-Hao, Won, M. and Hong, F. (2009) "Investigation of Settlement on A Joint Concrete Pavement" Journal of Performance of Constructed Facilities, ASCE. Volume 23, Number 6, pp. 440–446. December 2009.

39. Chen, Dar-Hao (2009) "Investigation of a Pavement Premature Failure on a Weak and Moisture Susceptible Base" Journal of Performance of Constructed Facilities, ASCE. Volume 23, Number 5, pp. 309–313. September/October 2009.

40. Chen, Dar-Hao, Won, M, Zhang, Q, and Scullion, T. (2009) "Field Evaluations of the Patch Materials for Partial Depth Repairs" Journal of Materials in Civil Engineering, ASCE September 2009, Vol. 21, Issue 9. pp 518–522.

41. Hong, F., and Chen, Dar-Hao (2009) "Effects of Surface Preparation, Thickness, and Material on Asphalt Pavement Overlay Transverse Crack Propagation". Volume 36, Number 9, pp. 1411–1420. September 2009. Canadian Journal of Civil Engineering.

42. Chen, Dar-Hao, Si, Z. and Saribudak, M. (2009) "Roadway Heaving Caused by High Organic Matter" Journal of Performance of Constructed Facilities, ASCE. Vol. 23, No. 2, pp. 100–108. March/April 2009.

43. Chen, Dar-Hao and Hong, F. (2009) "Field Verification of Smoothness Requirements for Weigh-In-Motion Approaches" Journal of Testing and Evaluation, ASTM. Vol. 37, No. 1 pp. 40–47. January 2009.

44. Hong, F. and Chen, Dar-Hao (2009) "Calibrating Mechanistic-Empirical Design Guide Permanent Deformation Models Based on Accelerated Pavement Testing" Journal of Testing and Evaluation, ASTM. Vol. 37, No. 1. pp. 31–39. January 2009.

45. Chen, Dar-Hao, Suh, C., and Won, M (2009) "Lessons Learned From Field and Laboratory Testing of a Dowel Bar Retrofit (DBR) Project". Journal of Performance of Constructed Facilities, ASCE. Vol. 23, No. 3. pp. 175–180. May/June 2009.

46. Chen, Dar-Hao (2008). "Field Experiences With the RDD and Overlay Tester for Concrete Pavement Rehabilitation" Journal of Transportation Engineering, ASCE. Volume 134, Number 1, pp. 24–33 January 2008.

47. Chen, Dar-Hao, and Scullion, T. (2008) "Forensic Investigations of Roadway Pavement Failures" Journal of Performance of Constructed Facilities, ASCE. Volume 22, Number 1, pp. 35–44. January/February 2008.

48. Chen, Dar-Hao, Won, M and Zha, X. (2008) "Performance of Dowel Bar Retrofit (DBR) Projects in Texas" Journal of Performance of Constructed Facilities, ASCE. Vol. 22, No. 3, pp. 162–170. May/June 2008.

49. Chen, Dar-Hao, and Scullion, T. (2008) "Detecting Subsurface Voids Using Ground- Coupled Penetrating Radar" Geotechnical Testing Journal, ASTM. Vol. 31. Number 3. pp. 217–224. May 2008.
50. Chen, Dar-Hao, Scullion, T., Tzen-Chin Lee, and Bilyeu, J. (2008) "Results from a Forensic Investigation of a Failed Cement Treated Base" Journal of Performance of Constructed Facilities, ASCE. Vol. 22, No. 3, pp. 143–153. May/June 2008.
51. Chen, Dar-Hao and Won, M (2008) "Field Performance Monitoring of Repair Treatments on Joint Concrete Pavements" Journal of Testing and Evaluation, ASTM Volume 36, Number 2. pp. 119–127. March 2008.
52. Chen, Dar-Hao, Li, Z. (2008) "Comparison of the Computational Methods for Transverse Profiles" Journal of Testing and Evaluation, ASTM. Vol. 36. Number 5. pp 473–480. September 2008.
53. Chen, Dar-Hao (2008) "Forensic Investigation of Mechanically Stabilized Earth (MSE) Approaches" Journal of Testing and Evaluation, ASTM. Vol. 36. Number 5. pp 443–452. September 2008.
54. Chen, Dar-Hao, Nam, B.H and Stokoe, K. (2008) "Application of The Rolling Dynamic Deflectometer to Forensic Studies and Pavement Rehabilitation projects" Transportation Research Record: Journal of the Transportation Research Board, No. 2084, pp. 73–82. Transportation Research Board of the National Academies, Washington, D.C., 2008.
55. Chen, Dar-Hao, Huang, Q.L. and Ling, J.M. (2008) "Shanghai's Experience on Utilizing the Rubblization For Jointed Concrete Pavement Rehabilitation" Journal of Performance of Constructed Facilities, ASCE. Vol. 22, No. 6, pp. 398–407. November/December 2008.
56. Chen, T.T., Chang, J.R., and Chen, Dar-Hao (2008) "Applying Data Mining Technique to Compute Load Damage Exponent for Rutting Through Full Scale Accelerated Pavement Testing". International Journal of Road Materials and Pavement Design. Vol. 9, No. 2, pp. 227–246.

57. Chen, Dar-Hao (2007). "Using Rolling Dynamic Deflectometer and Overlay Tester to Determine the Reflective Cracking Potential" Journal of Testing and Evaluation, ASTM. Volume 35, Number 6. pp. 644-654. November 2007.

58. Chen, Dar-Hao, and Won, M. (2007) "Field Investigations of Cracking on Concrete Pavements" Journal of Performance of Constructed Facilities, ASCE. Volume 21, Number 6, pp. 450–458. November/December 2007.

59. Chen, Dar-Hao, Zhou, F. Lee, L., Hu, S., Stokoe, K., and Yang, J. (2007) "Threshold Values for Reflective Cracking Based on Continuous Deflection Measurements" Canadian Journal of Civil Engineering. 34 (10): pp. 1257–1266. October 2007.

60. Chen, Dar-Hao, Nazarian, S. and Bilyeu, J. (2007) "Failure Analysis of a Bridge Embankment With Cracked Approach Slabs and Leaking Sand". Journal of Performance of Constructed Facilities, ASCE. Vol. 21. Number 5. pp 375–381. October 2007.

61. Chen, Dar-Hao, Zhou, F., and Yuan, J.B. (袁剑波) (2007) "Verification and Calibration of Vesys5w Fatigue Cracking Model Using Results from Accelerated Pavement Testing" Journal of Testing and Evaluation, ASTM. Volume 35, Number 5. pp. 544–552. September 2007.

62. Chen, Dar-Hao (2007). "Field and Lab Investigations of Prematurely Cracking Pavements" Journal of Performance of Constructed Facilities, ASCE. Vol. 21. Number 4. pp. 293–301. August 2007.

63. Chen, Dar-Hao, and Scullion, T. (2007) "Using Nondestructive Testing Technologies to Assist in Selecting the Optimal Pavement Rehabilitation Strategy. Journal of Testing and Evaluation, ASTM. Volume 35, Number 2 pp. 211–219. March 2007.

64. Zhou, F., Hu, S., Chen, Dar-Hao, and Scullion, T. (2007) "Overlay Tester: A Simple Performance Test for Fatigue Cracking" Transportation Research Board. Journal of Transportation Research Record No. 2001. pp 1–8.

65. Chen, Dar-Hao, Scullion, T., and Bilyeu, J. (2006) "Lessons Learned on Jointed Concrete Pavement Rehabilitation Strategies in Texas" Journal of Transportation Engineering ASCE. Vol. 132. Number 3. pp. 257–265. March 2006.

66. Chang, J.R., Hung, C.T., Chen, Dar-Hao (2006) "Application of an Artificial Neural Network on Depth to Bedrock Prediction" International Journal of Computational Intelligence. Volume-2, Issue-1. March 2006.

67. Chen, Dar-Hao, Bilyeu, J. Tom Scullion, Nazarian, S. and Chiu, C.T. (2006) "Failure Investigation of A Foamed Asphalt Project" Journal of Infrastructure Systems. ASCE. Vol. 12. Number 1. pp. 33–40. March 2006.

68. Chen, Dar-Hao, Zhou, F., Cortez, E. (2006) "Determination of Load Damage Relationships through Accelerated Pavement Testing" Journal of Testing and Evaluation, ASTM. Volume 34, Issue 4. pp. 312–318. July 2006.

69. Chen, Dar-Hao, **Chen, T.,** Scullion, T., and Bilyeu, J. (2006) "Integration of Field and Laboratory Testing to Determine the Causes of A Premature Pavement Failure" Canadian Journal of Civil Engineering. Volume 33, Number 11. pp. 1345–1358. November 2006.

70. Chen, Dar-Hao, Lin, D-F, Liau, P-H, and Bilyeu, J. (2005) "A Correlation Between Dynamic Cone Penetrometer Values and Pavement Layer Moduli," Geotechnical Testing Journal, ASTM. Vol. 28. Number 1. pp. 42–49. January 2005.

71. Chen, Dar-Hao, Harris, Pat, Scullion, T., and Bilyeu, J. (2005) "Forensic Investigation of a Sulfate-Heaved Project in Texas. Journal of Performance of Constructed Facilities, ASCE. Vol. 19. Number 4. pp 324–330. November 2005.

72. Chang, J.R., Chen, Dar-Hao, and Hung, C.T. (2005) "Select Preventive Maintenance Treatments Using TOPSIS for SPS-3 Sites in Texas". Transportation Research Board. Journal of Transportation Research Record. No. 1933, pp. 62–71.

73. Lee, L., Stokoe, K., Chen, Dar-Hao, and Nam, B.H. (2005) "Monitoring Pavement Changes in a Rehabilitation Project with Continuous RDD

Profiles". Transportation Research Board. Journal of Transportation Research Record. No. 1905, pp 3–16.

74. Chang, J.R., Kuan-Tsung Chang, and Chen, Dar-Hao (2006) "Application of 3D Laser Scanning on Measuring Pavement Roughness," Journal of Testing and Evaluation, ASTM, Vol. 34, No. 2. pp 1–9. Mar. 2006.

75. Chen, Dar-Hao, Scullion, T., Bilyeu, J., and Won, M. (2005) "A Detailed Forensic Investigation and Rehabilitation Recommendation on IH-30" Journal of Performance of Constructed Facilities, ASCE. Vol. 19. Number 2. pp 155–164. May 2005.

76. Chang, J.R., Hung, C.T. and Chen, Dar-Hao (2006) "Application of an Artificial Neural Network on Depth to Bedrock Prediction" International Journal of Computational Intelligence Research. ISSN 0973-1873 Vol. 2, No. 1 (2006), pp. 33–39.

77. Lee, J. L., Chen, Dar-Hao, and Stokoe, K. (2004) "Evaluating the Potential for Reflection Cracking Using the Rolling Dynamic Deflectometer" Transportation Research Board. Journal of Transportation Research Record No. 1869. pp. 16–24. 2004.

78. Chen, Dar-Hao, Lin. D-F, and Luo, H-L (2003) "Effectiveness of Preventative Maintenance Treatments Using Fourteen SPS-3 Sites in Texas," Journal of Performance of Constructed Facilities, ASCE. Vol. 17. Number 3. pp 136–143. August 2003.

79. Chen, Dar-Hao, Bilyeu, J. Scullion, T., Zhou, F., and Lin, D.F. (2003) "Forensic Evaluation of the Premature Failures of the Texas SPS-1 Sections," Journal of Performance of Constructed Facilities, ASCE. Vol. 17. Number 2. pp 67–74. May 2003.

80. Chen, Dar-Hao, Scullion, T., Bilyeu, J., Yuan, D., and Nazarian, S. (2002) "Forensic Study of a Warranty Project On US82," Journal of Performance of Constructed Facilities, ASCE. Vol. 16. Number 1. pp 21–32. Feb. 2002.

81. Chen, Dar-Hao, and Hugo, H. (2001) "Comparison of the Effectiveness of Two Pavement Rehabilitation Strategies," Journal of Transportation Engineering ASCE. Jan./Feb. 2001 Vol. 127 No. 1. pp 47–58.

82. Zhou, F. J., Scullion, T., and Chen, Dar-Hao (2002) "Laboratory Characterization of Asphalt Mixes of SPS-1 Sections on US281," International Journal on Road Materials and Pavement Design. Vol. 3-No.4. pp. 439–454. 2002.

83. Chen, Dar-Hao, Bilyeu, J., Lin, H, and Murphy, M. (2000) "Temperature Correction on FWD Measurements" Transportation Research Board. Journal of Transportation Research Record No. 1716. pp. 30–39. 2000.

84. Chen, Dar-Hao, Bilyeu, J., Debbie Walker, and Mike Murphy (2001) "Study of Rut-Depth Measurements" Transportation Research Board. Journal of Transportation Research Record No. 1764. pp. 78–88. 2001.

85. Chen, Dar-Hao, Jian-Neng Wang and Bilyeu, J. (2001) "Application of the DCP in Evaluation of Base and Subgrade Layers" Transportation Research Board. Journal of Transportation Research Record No. 1764. pp. 1–10. 2001.

86. Chen, Dar-Hao (1999) "Determination of Bedrock Depth from FWD Data," Transportation Research Board. Transportation Research Record No. 1655. pp 127–134. 1999.

87. Stokoe K., Bay, J., Rosenbald, B., Murphy, M., Fults, K, and Chen, Dar-Hao (2000) "Super-Accelerated Testing of a Flexible Pavement with the Stationary Dynamic Deflectometer (SDD)" Transportation Research Board. Journal of Transportation Research Record No. 1716. pp. 98–107.

88. Chen, Dar-Hao, Zaman, M. M., and Laguros, J.G. (1995) "Characterization of Base/Subbase Materials under Repetitive Loading," Journal of Testing and Evaluation ASTM, Vol. 23, No. 3, pp. 180–188. May 1995.

89. Chen, Dar-Hao, Zaman, M. M., and Laguros, J.G. (1995) "Assessment of Computer Programs for Analysis of Flexible Pavement Structure," Transportation Research Board. Transportation Research Record No. 1482. pp 123–133. 1995.

90. Chen, Dar-Hao, Zaman, M. M., and Laguros, J.G. (1994) "Reliability of Resilient Moduli for Aggregate Materials: Variability Due to Testing Procedure and Aggregate Type," Transportation

Research Board. Journal of Transportation Research Record No. 1462. pp 57–64. 1994.

91. Yuan, D., Nazarian, S., Chen, Dar-Hao and Hugo, H. (1998) "Use of Seismic Pavement Analyzer to Monitor Degradation of Flexible Pavements under Texas Mobile Load Simulator," Transportation Research Board. Journal of Transportation Research Record No. 1615. pp. 3–10. 1998.

92. Chen, Dar-Hao and Hugo, H (1998) "Test Results and Analyses of the Full-Scale Accelerated Pavement Testing of TxMLS," Journal of Transportation Engineering ASCE. Vol. 124 No. 5. pp 479–490. Sep./Oct. 1998.

93. Chen, Dar-Hao, Fults, K., and Murphy, M. (1997) "The Primary Results for the First TxMLS Test Pad Transportation Research Board. Journal of Transportation Research Record No. 1570. pp. 30–38. 1997.

94. Zaman, M. M., Chen, Dar-Hao, and Laguros, J.G. (1994) "Resilient Modulus Testing of Granular Material and Their Correlations with Other Engineering Properties," Journal of Transportation Engineering ASCE. Vol. 120, No. 6, pp. 967–988. Nov./Dec. 1994.

95. Chen, Dar-Hao (1999) "Pavement Distress Under Accelerated Trafficking," Transportation Research Board. Journal of Transportation Research Record No. 1639. pp 120–129. 1999.

96. Chen, Dar-Hao, Fernando, E., and Murphy, M. (1996) "Application of Falling Weight Deflectometer Data for Analysis of Superheavy Loads," Transportation Research Board. Journal of Transportation Research Record No. 1540. pp 83–90. 1996

Published Papers (EI Indexed)

1. Abu Faruk and Dar Hao Chen (2016). "Traffic Volume and Load Data Measurement Using a Portable Weigh in Motion System: A Case Study" International Journal of Pavement Research and Technology.

2. Dar Hao Chen (2015). "Pavement Research Programs and Texas's Experience" International Journal of Pavement Research and Technology. Vol 8, Issue 2. pp IV–VII. March-2015.
3. Dar Hao Chen. "Updates on Two TxDOT's Research Implementation Projects" International Journal of Pavement Research and Technology Vol.8 No.2 March 2015.
4. Dar Hao Chen, Huang-Hsiung Lin, Jing Rong Zou, and Zhao Bin Xie (2014) "Characterization of Trackless Tacks Using Fracture Mechanics". ASCE Geotechnical Special Publications. Number 246. pp 186–193. (EI Indexed).
5. Dar Hao Chen, Wen Yi, Peng Ying, and Kei Fei Liu (2014) "Nanotechnology on Trackless Tacks". ASCE Geotechnical Special Publication Number 244. pp 17–23.(EI Indexed).
6. Dar Hao Chen and Li Kun (2014) "Using Electromagnetic Waves to Detect Subsurface Voids". ASCE Geotechnical Special Publication Number 252. pp 32–38. (EI Indexed).
7. Dar Hao Chen and Li Kun (2014) "Improving Surface Characteristics by Using Diamond Grinding" ASCE Geotechnical Special Publication Number 249. pp 83–88. **(EI Indexed)**.
8. ANM Faruk, Chen, Dar-Hao, C Mushota, M Muya, LF Walubita (2014) "Application of Nano-Technology in Pavement Engineering: A Literature Review" ASCE Geotechnical Special Publication Number 244. pp 9–16. **(EI Indexed)**.
9. Chen, Dar-Hao, and Lee, J. (2011) "Utilizing Ground Penetrating Radar for Roadway Structure Inspections" ASCE Geotechnical Special Publications. Number 215. pp 8–15. **(EI Indexed)**.
10. Chen, Dar-Hao, Huang, J. and Lee, J. (2011) "Pavement Premature Failure on a Moisture Susceptible Base" ASCE Geotechnical Special Publications. Number 222. pp 305–312. **(EI Indexed)**.
11. Chen, Dar-Hao, Fowler, D., Whitney, D., and Won, M. (2011) "Results of Repairs on Texas Longitudinal Joints and Cracks" ASCE Geotechnical Special Publications. Number 218. pp 208–214. **(EI Indexed)**.

12. Chen, Dar-Hao, Li, J., and Jun Xie (2011) "Performance of Fiber Reinforced Polymer Patching Binder for Minimizing Reflective Cracking" ASCE Geotechnical Special Publications. Number 212. pp. 175–182. (**EI Indexed**).

13. Chen, Dar-Hao, Xie, J., and Scullion, T. (2011) "Using GPR and FWD to Assist in Selecting the Optimal pavement Rehabilitation Strategy" ASCE Geotechnical Special Publications. Number 212. pp 63–70. (**EI Indexed**).

14. Zahid Hossain, Musharraf Zaman, Edgar A. O'Rear, and Chen, Dar-Hao (2011) Effectiveness of Advera® in Warm Mix Asphalt. ASCE Geotechnical Special Publications Number 218. pp. 9–16. (**EI Indexed**).

15. Nam, B., Lee, J., Stokoe, K., Chen, Dar-Hao, and Jung-Su Lee (2011) "Application of RDD Continuous Profiling in Maintenance and Rehabilitation of Concrete Pavements" ASCE Geotechnical Special Publications. ASCE Geotechnical Special Publications. Number 218. pp 125–132. (**EI Indexed**).

16. Hongyuan Fu, Guiyao Wang, Jianjun Xu, Xiang Xin, Xudong Zha, and Dar-Hao Chen (2010) "Modeling Differential Settlement in the Partial-Cut and Partial-Fill Embankments of the Mountainous Expressways of China" ASCE Geotechnical Special Publications Number 199. pp. 1806–1815. (**EI Indexed**).

17. Chen, Dar-Hao, Wimsatt, A., and Bilyeu, J. (2009) "Using Ground Penetration Radar Techniques for Roadway Structure Safety Evaluation" ASCE Geotechnical Special Publications Number 189. pp 116–122. (**EI Indexed**).

18. Scullion, T., and Chen, Dar-Hao (2009) "A Key Tool for Directing Future Research" ASCE Geotechnical Special Publications Number 191. pp 87–95. (**EI Indexed**).

19. Chen, Dar-Hao, Crawford, T., Fowler, D. Jirsa, J., Stringer, M., Whitney, D. and Won, M. (2009) "Repair of Longitudinal Joints and Cracks" ASCE Geotechnical Special Publications Number 196. pp 119–124. (**EI Indexed**).

20. Chen, Dar-Hao, Bilyeu, J., and Li, Z. (2009) "Field Evaluation of Damages from Super Heavy Load Moves" ASCE Geotechnical Special Publications Number 193. pp 187–192. **(EI Indexed)**

21. Chen, Dar-Hao, Bilyeu, J. and, Chang, J.R. (2005) "A Review of the Superheavy Load Permitting Program in Texas" International Journal of Pavement Engineering. Vol. 6 (1) June 2005, pp. 45–53. **(EI Indexed).**

22. Chen, Dar-Hao, Scullion, T., Bilyeu, J., Fults, K. and Murphy, M. (2004) "Pavement Forensics: Investigating Failure" Better Roads. pp 36–40. January 2004. **(EI Indexed)**

23. Chiu, Chui-Te, Lee, M-G, and Chen, Dar-Hao (2003) "Study of Profile Measurements Using Six Different Devices" Constructing Smooth Hot Mix Asphalt (HMA) Pavements, ASTM STP 1433, P102–113. American Society for Testing and Materials, West Conshohocken, PA, 2003. **(EI Indexed)**

24. Zhou, F., Chen, Dar-Hao, Scullion, T. and Bilyeu, J. (2003) "Case Study: Evaluation of Laboratory Test Methods to Characterize Permanent Deformation Properties of Asphalt Mixes" International Journal of Pavement Engineering. Vol. 4 (3) September 2003, pp. 155–164. **(EI Indexed)**

25. Chen, Dar-Hao, Lin. D-F., and Bilyeu, J. (2002) "Determination of the Effectiveness of Preventative Maintenance Treatments," International Journal of Pavement Engineering. Vol. 3 (2), pp. 71–83. 2002. **(EI Indexed)**

26. Chang, J.R. Lin, Jyh-Dong, Chung, W.C. and Chen, Dar-Hao (2002) "Evaluating the Structural Strength of Flexible Pavements in Taiwan Using the Falling Weight Deflectometer" International Journal of Pavement Engineering, Vol. 3, No. 3, pp. 131–141. 2002. **(EI Indexed)**

27. Dar-Hao Chen and Bilyeu, J. (2001) "Assessment of a Hot-in-Place Recycling Process" Tamkang Journal of Science and Engineering, Vol. 4, No. 4, pp. 265–276 (2001) 265. **(EI Indexed).**

28. McDaniel, M., Yuan, D, Chen, D-H and Nazarian, S. (2000) "Use of Seismic Pavement Analyzer in Forensic Studies in Texas", Nondestructive Testing

of Pavements and Backcalculation of Moduli: Third Volume, ASTM STP 1375, p346–364. S.D. American Society for Testing and Materials, 2000. (**EI Indexed**)

29. Chen, Dar-Hao, Lin, H and Hugo, H (2000) "Application of VESYS3AM in Characterization of Permanent Deformation" International Journal of Pavement Engineering, Vol. 1 (3), pp. 171–192. 2000. (**EI Indexed**)

30. Chen, Dar-Hao, Wu, W., He, R., Bilyeu, J., and Arrelano, M. (1999) "Evaluation of In-Situ Resilient Modulus Testing Techniques," Geotechnical Special Publication No. 89 ASCE. pp 1–11. 1999. (**EI Indexed**)

31. Najjar, Y.N., Attoh-Okine, N.O., Basheer, I.A. and Chen, Dar-Hao et al. (1999) "Use of Artificial Neural Networks in Geomechanical and Pavement Systems" Transportation Research Board. Transportation Research Circular. E-C012, December 1999. (**EI Indexed**)

32. Chen, Dar-Hao, and Bilyeu, J., Hugo, H (1999) "Monitoring Pavement Response and Performance Using In-Situ Instrumentation Under Full-Scale Accelerated Loading," Field Instrumentation for Soil and Rock, ASTM STP 1358, p121–134. G. N. Durham and W. A. Marr, Eds., American Society for Testing and Materials, 1999. (**EI Indexed**)

33. Chen, Dar-Hao, Murphy, M., and Yeggoni, M. (1996) "Application of FWD in Analyzing Finite Width Effect of Pavements," Proceeding, v 2, p1018–1021. 11th ASCE Engineering Mechanics Conference, Boca Ration, Fla., May 19–22, 1996. (**EI Indexed**).

34. Chen, Dar-Hao, and Murphy, M. (1996) "Analysis of Pavement Structural Responses Using In-Situ Instrumentation," Proceeding, v 2, p705–708. 11th ASCE Engineering Mechanics Conference, Boca Ration, Fla., May 19–22, 1996. (**EI Indexed**)

35. Chen, Dar-Hao, Zaman, M. M., and Laguros, J.G. (1994) "Assessment of Distress Models for Prediction of Pavement Service Life," Proceedings

of the Third Materials Engineering Conference. n 804, Proceeding p1073–1080. New Materials and Methods for Repair. San Diego, California. Nov. 14–16, 1994 **(EI Indexed).**

36. Chen, Dar-Hao, Zaman, M. M., and Kukreti, A. R. (1993) "Laboratory Testing and Constitutive Modeling of Coal Including Anisotropy," Material Research Society Symposium. Structure and Properties of Energetic Material. Vol. 296. p 349–354. 1993. **(EI Indexed).**

核心期刊
黄琴龙, 陈达豪, 凌建明, 徐柱杰, 共振碎石化在上海水泥混凝土路面改建中的成效-长沙理工大学学报 (自然科学版) 2008年 第5卷 第02期. ISSN : 1672-9331(2008)02-0103-04

Refereed Conference Proceedings

1. Zhou, F., Hu, S., Chen, Dar-Hao, and Scullion, T. "RDD Data Interpretation with A Focus on "Critical" Sections and Joints of Concrete Pavements", 11th International Conference on Asphalt Pavement, Nagoya, Japan, Aug. 1–6, 2010.

2. Chen, Dar-Hao, Won, M., and Scullion, T. (2009) "Minimizing Reflective Cracking with Applications of the Rolling Dynamic Deflectometer & Overlay Tester" Proceedings of the National Conference on Preservation, Repair, and Rehabilitation of Concrete Pavements. pp 3–14. St. Louis, Missouri. April 22–24, 2009.

3. Zhou, F., Hu, S., Scullion, T., Chen, Dar-Hao, and Qi, X. (2007) "Development and Verification of the Overlay Tester Based Fatigue Cracking Prediction Approach" Journal of the Association of Asphalt Paving Technologists. Volumes 76. pp 627–662.

4. Chen, Dar-Hao (2005) "Lessons Learned from LTPP and Several Recycled Sections in Texas," Transportation Research Board. Transportation Research Circular. E-C078, pp 70–86. October 2005.

5. Stokoe, K. H., Lee, J., Chen, Dar-Hao, Oshinski, E., Joh, S. H., Nam, B. H. (2005) "Continuous Deflection Measurements of Highway and Airport Pavements with the Rolling Dynamic Deflectometer", Proceedings of the 5th International Conference on Road & Airfield Pavement Technology, May 10–12, 2005, Seoul, Korea.

6. Chen, Dar-Hao, Wang, W., Zhou, F. Kenis, B., Nazarian, S. and Scullion, T. (2004) "Rutting Prediction Using Calibrated Model with TxMLS and AASHO Road Data" Proceeding CD. 2nd International Conference on Accelerated Pavement Testing. September 26–29, 2004. Minneapolis, Minnesota.

7. Chen, Dar Hao, Scullion, T., Bilyeu, J., Fults, K., Murphy, M. (2003) "Forensic Studies in Texas" Proceeding, International Conference on Highway Pavement Data, Analysis and Mechanistic Design Application. Sep. 7–10, 2003 Columbus, Ohio.

8. Lee, J.L., Turner, D.J., Chen, Dar Hao, Bilyeu, J., and Stokoe, K.H. (2003) "Project-Level Study Using Continuous Deflection Profiles Measured With the Rolling Dynamic Deflectometer. Proceedings of the International Conference on Highway Pavement Data, Analysis and Mechanistic Design Applications, September 7–10, 2003, Columbus, Ohio, Volume 2 (2003).

9. Chen, Dar-Hao, and Lin, H. (1999) "Predictive Equation for Permanent Deformation," Proceeding, CD-ROM, 1st International Conference on Accelerated Pavement Testing. Reno, Nevada. October 18–20, 1999.

10. Hugo, F., Chen, Dar-Hao, Fults, K., Smit A., Bilyeu, J. (1999) "An Overview of the TxMLS Program and Lessons Learned," Proceeding, CD-ROM, 1st International Conference on Accelerated Pavement Testing. Reno, Nevada. October 18–20, 1999.

11. Nazarian, S., Yuan, D., Chen, Dar-Hao, and McDaniel, M. (1999) "Use of Seismic Methods in Monitoring Pavement Deterioration During Accelerated Pavement Testing with TxMLS," Proceeding, CD-ROM, 1st International

Conference on Accelerated Pavement Testing. Reno, Nevada. October 18–20, 1999.

12. Chen, Dar-Hao, Murphy, C, Pilson, C., and Hudson, R. (1997) "Testing and Analysis of the TxMLS Test Pads At Victoria, Texas, "Proceeding, p1205–1224. 8th International Conference on Asphalt Pavements, Seattle, Washington, August 10–14, 1997.

13. Chen, Dar-Hao, Bilyeu, J., and Murphy, M. (1998) "The Effect of Bedrock Depth on Pavement Response, "Proceeding. pp 399–407. 49th Highway Geology Symposium, Arizona, September 10-14, 1998.

14. Chen, Dar-Hao (1998) "Determination of the Effectiveness of the Pavement Rehabilitation Strategy, "Proceeding, CD-ROM. 1998 Science, Engineering & Technology Seminars (SETS). Houston, TX. May 23–24. 1998.

15. Zaman, M. M., Chen, Dar-Hao, and Laguros, J.G. (1995) "Cyclic Triaxial Testing of Granular Materials for Application in Pavement Design," Proceeding Vol. 1, p53–56. 3rd International Conference on Recent Advances in Geotechnical Earthquake Engineering and Soil Dynamics, St. Louis, Missouri. April 2–7, 1995.

16. Chen, Dar-Hao, Zaman, M. M., Laguros, J.G., and Senkowski, L. (1995) "Numerical Modeling for Assessment of Cooling Rates of Hot Mix Asphalt Overlays," Proceeding, p105–108. Sixth International Conference on Civil and Structural Engineering Computing, Cambridge, England, August 28–30, 1995.

17. Chen, Dar-Hao, Zaman, M. M., and Laguros, J.G. (1995) "Mechanistic-Empirical Methodology for Conventional Flexible Pavement Design," Proceeding, p191–194. Sixth International Conference on Civil and Structural Engineering Computing, Cambridge, England, August 28–30, 1995.

18. Chen, Dar-Hao, Zaman, M. M., Mehta, Y., and Laguros, J.G. (1995) "Using a Finite Element Program ABAQUS in Predicting the Cooling Rates of Hot Mix Asphalt," Proceeding, p663–668. Numerical Models in Geomechanics V, Davos, Switzerland, September 6–9, 1995.

19. Najjar,Y. M., Chen, Dar-Hao, and Zaman, M. M. (1993) "Application of Nonlinear Finite Element Method in Prediction of Ground Subsidence Due to Underground Mining," Proceeding, p140–149. 1st Canadian Symposium on Numerical Applications in Mining and Geomechanics, Montreal, Quebec, Canada, March 27–30, 1993.

20. Chen, Dar-Hao, Zaman, M. M., and Laguros, J.G. (1993) "Dynamic Testing of Aggregate Bases," Proceeding, p259–274. 6th International Conf. on Soil Dynamics and Earthquake Engineering, Bath, U.K., June 14–16, 1993.

21. Najjar,Y. M., Zaman, M. M., and Chen, Dar-Hao (1993) "Anisotropic Constitutive Model for Pressure Sensitive Materials," Proceeding 4th Int'l Symp. on Plasticity. Maryland. July 19–23, 1993.

22. Zaman, M. M., Kukreti, A. R., and Chen, Dar-Hao (1993) "Numerical Modeling of Mining Induced Subsidence," Proceeding Vol. 1. p521–526. 2nd Asian-Pacific Conference on Computational Mechanics Sydney, Australia. Aug. 3–6, 1993.

23. Chen, Dar-Hao, Zaman, M. M., and Oyesanya, A. (1994) "3-D Testing and Constitutive Modeling of Inherently Anisotropic Sand," Proceeding p400–414. Seventeenth Southeastern Conference on Theoretical and Applied Mechanics, Hot Springs National Park, Arkansas. April 10–12, 1994.

24. Chen, Dar-Hao, McDaniel, M., and Graham, G. (1996) "Study of FWD Deflection and Strain and Pressure Measurement for Accelerated Pavement Testing," Proceeding, p204–213. Texas Section ASCE Meeting September 18–20, 1996.

25. Chen, Dar-Hao, Meilahn, Nancy, Mike Murphy, and Fults, K. (1997) "Evaluation of the Change of Material Properties Under Accelerated Trafficking of Texas Mobile Load Simulator," Proceeding, CD-ROM, XIIIth International Road Federation (IRF) World Meeting. Toronto 1997.

26. Chen, Dar-Hao, and Fults, K. (1997) "Texas Mobile Load Simulator (TxMLS)" Proceeding. B2-3-1 -B2-3-11 International Center for Aggregates Research 5th Annual Symposium. April 21–23, 1997. Austin, Texas.

27. Zaman, M. M., Chen, Dar-Hao, and Najjar, Y. M. (1997) "Multiaxial Testing and Constitutive Modeling of Coal" Proceedings of 9th International Conference on Computer Methods and Advances in Geomechanics. p285–292. Wuhan, China. November 2–7, 1997.

28. Chen, Dar-Hao, Bilyeu, J. and Murphy, M. (2001) "Stiffness Evaluation of Chemical Stabilizers used in the Dallas and Austin Districts," Proceedings CD-ROM. Second International Symposium on Maintenance and Rehabilitation of Pavements and Technological Control. Auburn, Alabama, July 29 – Aug. 1, 2001.

29. Chen, Dar-Hao, Murphy, M and., Fults, K. (1998) "Accelerated Pavement Testing With TxMLS", Proceeding Vol. 1. p478–486. 3rd International Conference On Road & Airfield Pavement Technology. Beijing, China. 1998.

Investigation on the Use of Crumb Rubber and Bagasse Ash in Road Construction

M. Patil Akshay[1], Anand B. Tapase[2(✉)], Y. M. Ghugal[3],
B. A. Konnur[1], and Shrikant Dombe[1]

[1] Department of Civil Engineering, Government College of Engineering, Karad,
Karad, Maharashtra, India
patil.akshay1595@gmail.com, bakonnur@gmail.com,
shrikantdombe@gmail.com

[2] Department of Civil Engineering, Rayat Shikshan Sanstha's Karmaveer
Bhaurao Patil College of Engineering, Satara, Maharashtra, India
tapaseanand@gmail.com

[3] Applied Mechanics Department, Government College of Engineering, Karad,
Karad, Maharashtra, India
yuwraj.ghugal@gcekarad.ac.in

Abstract. The naturally occurring materials like aggregates, sand, bitumen, etc. are depleting day by day due to its extensive usage in various civil infrastructural and building construction projects. On the other hand, the disposal of industrial and domestic waste in an eco-friendly way is a thrust area for researchers and has ample scope for its utilization as an alternative to the conventional exhausting natural materials. A number of researchers have already reported the potential use of various waste materials in road construction. The present study is an attempt to report the obtained results from extensive literature review upon the use of waste materials like tyre rubber, bagasse ash, in road construction. Furthermore, as a sustainable construction method, the experimental investigation on studying the effects of adding scrap rubber and bagasse ash as a partial replacement to bitumen and filler in the bituminous mixes respectively done by using the Marshall stability testing machine. The results obtained from the comparative study showed that scrap rubber can be used in road construction as a partial alternative to bitumen but the additions of bagasse ash lower done the strength & quality of the bituminous mix. So it is concluded that scrap rubber can be effectively used in the road construction as an alternative to bitumen but the use of bagasse ash is not been recommended in bituminous layers.

Keywords: Scrap rubber · Partial replacement · Bagasse-ash

1 Introduction

As the amount of cars in India is increasing quickly, tire waste becomes a significant environmental fear. Crumb rubber is a material generated when used tires are destroyed and commuted. Nearly 60% of waste tyres current are disposed of through urban and rural regions. This creates numerous conservation problems, including air pollution

S. Badawy and D.-H. Chen (Eds.): GeoMEast 2019, SUCI, pp. 1–12, 2020.
https://doi.org/10.1007/978-3-030-34196-1_1

(owing to the burning of tyres) and artistic pollution that causes serious health-related problems. These are abiotic, disposable goods that cause environmental pollution to these materials. This creates numerous conservation problems, including air pollution (owing to the burning of tyres) and artistic pollution that causes serious health-related problems. These are abiotic, disposable goods that cause environmental pollution to these materials. In contemporary years, rubber waste by-products are being used with excellent concern in many developing nations in highway building. Practical, commercial and environmental criteria are used to select these materials in road construction. Millions of tons of rubber waste are manufactured in India every year. Using these materials in road construction can effectively reduce the issues of pollution and disposal. Caring for the bulk trick of these waste in India, it was considered necessary to examine these products and create requirements to boost the use of rubber waste in highway manufacturing, where greater financial returns could be feasible. These materials should be used in road construction in each and every part of our country.

The use of crumb rubber, which is the reprocessed tire rubber, as an additive in asphalt mixture is considered a sustainable construction method. The addition of crumb rubber to the bitumen binder enhanced the physical properties of rubberized bitumen binder as indicated by the reduction in penetration and ductility. There are two fundamental procedures in asphalt blend for adding crumb rubber, wet and dry procedures. Crumb rubber is added to asphalt in a wet method, allowing the reaction of rubber and asphalt. The primary method of the moist method is rubber swelling.

In the dr phase, before adding bitumen, Crumb rubber is blended with the warm aggregate. Adding tire rubber to asphalt mixtures using a dry method could enhance the characteristics of permanent deformation resistance at elevated temperatures and cracking at low temperatures. As specified, the rubberized asphalt mixture with the wet process could achieve the desired Volumetric parameters. The aim of this research was to explore the impact of using the wet method to add crumb rubber to asphalt mixtures.

Sugarcane bagasse ash is the sugarcane refinement's natural product. It is the fiber remaining to create juices after the method of extraction that comes from sugar cane. When the juice was extracted from sugarcane, the amounts of product volumes in sugar factories render it a waste product Sugarcane bagasse ash is the natural product of sugarcane refinement. After the extraction technique that comes from sugar cane, it is the fiber that remains to produce juices. The quantities of product quantities in sugar mills make it a waste product when the juice was extracted from sugar cane.

1.1 Dense Bituminous Macadam (DBM)

For use as a base course and/or binder course, the dense bituminous macadam (DBM) is currently specified. Two DBM gradations are specified in section 505 of the 2013 MORTH specifications: Grading 1 has an NMAS (nominal maximum aggregate size) of 37.5 mm and Grading 2 has an NMAS of 25 mm MORTH composition of DBM Grading 1 and 2. The specified percentage of fine aggregates is the same in both gradings (28–42%), the main difference is only some large aggregate particles (25–45 mm size) are the same in both gradations. The use of large stone mix (37.5 mm or bigger NMAS) has several drawbacks such as segregation and elevated permeability. These disadvantages outweigh the "marginal" gain instability if any; over a mixture of

25 mm NMAS. As Grading 1 is extremely permeable, it must be closed or overlaid before rainy season otherwise water will penetrate it and harm the fundamental course of WMM. experienced Indian highway technicians recommend this but the alternatives are merely to completely prohibit the problematic DBM Grading 1 and use only the DBM Grading 2. On many national highways in India deteriorated DBM grading 1 in the reduced lift of the complete DBM, which was disintegrated owing to drying, could not be recovered intact by coring. Based on the previous debate, in the flexible pavement, problematic DBM Grading 1 should not be used (Table 1).

Table 1. Grading dense bituminous macadam as per morth

Grading	1	2
Nominal aggregate size*	37.5 mm	26.5 mm
Layer thickness	75–100 mm	50–75 mm
IS Sieve 1 (mm)	Cumulative % by weight of total aggregate passing	
45	100	
37.5	95–100	100
26.5	63–93	90–100
19	–	71–95
13.2	55–75	56–80
9.5	–	–
4.75	38–54	38–54
2.36	28–42	28–42
1.18	–	–
0.6	–	–
0.3	7–21	7–21
0.15	–	–
0.075	2–8	2–8
Bitumen content % by mass of total mix	Min 4.0**	Min 4.5**

2 Aims and Objectives

To study the use of flexible pavement bagasse ash in DBM along with partial replacement of bitumen by Crumb rubber by wet mixing method. Finding an alternative method for eco-friendly disposal of Crumb rubber and bagasse ash. To find an appropriate alternative in flexible pavements to conventional materials with cost reduction & improvement in strength & other parameters. Using waste material in flexible paving without growing unit costs and without sacrificing durability. Using software to verify the suitability of waste products in the flexible pavement.

3   Materials and Methods

3.1   Bitumen

The bitumen used in the 60/70 grade of penetration in this inquiry. As per IS code (India), the bitumen used in the bitumen investigation characteristics was verified. Table 2 shows the properties. To determine the optimum content of asphalt, bitumen content as per minimum MoRTH is 4.5 specified table 500-10 as per grading 2 bitumen content was varied 4.0%, 4.25%, 4.5%, 4.75%, 5% by weight of asphalt blend. For each proportion, three samples are conducted too much. On the basis of the Marshall Stability Test, the optimum bitumen content was determined, corresponding requirement specification. Higher content of crumb rubber and bagasse ash had a powerful impact on the reduction of bitumen penetration and ductility.

Table 2. Physical properties of bitumen

Sr. no.	Tests conducted	Test results	Specifications	IS codes
1	Specific Gravity	1.02	0.97–1.02	IS-1202
2	Flashpoint, °C	270	220 °C	IS-1209
3	Penetration test	65	50–90	IS-1203
4	Softening point test, °C	49	Min 55 °C	IS-1205
5	Ductility test	51	Min 40	IS-1208

4   Crumb Rubber

The fast growth of the automotive sector and a greater standard of living of individuals in India, the number of cars raised significantly, India facing the environmental issue of large-scale waste tires disposal. How to handle the enormous amount of waste tires in India has become an urgent environmental issue. Every year, tens of millions of tires are discarded throughout the Middle East. Because tires have a long life and are non-biodegradable, disposal of waste tires is a challenge. The traditional technique of managing waste pneumatics was storage or illegal dumping or landfilling, all of which are a short-term solution.

The world's disposal of waste tires has three main methods to cope with such things as landfills, burning & recycling. Scrap tire (recycled tire rubber) applied to pavement can be the best way to reduce large quantities of waste tires while at the same time improving the engineering properties of asphalt mixtures.

Crumb rubber is a word used in automotive and truck scrap tires for recycled rubber. There are two significant technologies for mechanical processing of the ambient crumb rubber and cryogenic processing of the two procedures, the cryogenic method is more costly but it generates smoother and lower crumbs.

This article uses a moist method to present experimental studies on recycled asphalt mixtures modified by tire rubber. The advantages of wet alteration are the extensive and controllable interaction between rubber and bitumen that enables elevated binder quality to be obtained.

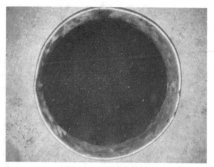

Waste tyre at Karad near NH4   Crumb rubber

4.1 Ambient Mechanical Grinding

The breaking up of a scrap tire occurs at or above ordinary room temperature during the ambient mechanical grinding process. Ambient grinding is a multi-step technology that utilizes whole or pre-treated shredded car or truck tires, or sidewalls or treads. Sequentially separate the rubbers, metals, and fabrics. Pneumatics pass through a shredder that breaks the tires into chips.

The chips are fed into a granulator which, while removing steel and fiber in the process, breaks them into tiny pieces. Any remaining steel and fiber are separated magnetically By combining shaking displays with wind seams. In secondary granulators and high-speed rotary mills finer rubber particles can be acquired by further grinding. Ambient grinding is the manufacturing method most crumb manufacturers are using. The machines most frequently used in ambient crops for fine grinding are:

(1) Secondary granulators
(2) High-speed
(3) rotary mills
(4) Cracker mills

4.2 Cryogenic Grinding

Cryogenic grinding refers to grinding scrap tires with liquid nitrogen or commercial refrigerants at temperatures near minus 80 °C. Cryogenic processing uses pretreated car or truck tires as feedstock, most often in the form of chips or granulates produced in the environment using fluid nitrogen or commercial refrigerants to embitter the rubber, processing takes place at very low temperatures. It can be a four-phase scheme that involves an original decrease in size, refrigeration, separation, and friction. The material enters a freezing chamber where liquid nitrogen is used to cool it from −80 to −120 °C, below the stage where rubber stops acting as a flexible material and can readily be crushed and broken.

In a hammer mill, fibers and metal are readily segregated due to their brittle state. The granulate then moves through a sequence of magnetic screens and seals to remove the last remains of impurities. This method needs less power than others, producing much finer quality rubber crumb.

Paravitasri Wulandari et al. (2016) demonstrated the usefulness of crumb rubber as an additive in asphalt concrete mixture, which is recycled tire rubber, as an additive in a warm blend of asphalt is regarded a sustainable method of building. Crumb rubber adding crumb rubber to the proportion tends to boost asphalt mixture strength and quality. Modified blend required less asphalt content to improve stability and reduce inflow crumb rubber.

N. S. Mashaan et al. evaluated the addition of crumb rubber to the bitumen binder to enhance the physical characteristics of the rubberized bitumen binder as demonstrated by decreased penetration and ductility and increased elastic recovery, thereby enhancing the elasticity of the rubberized binder and enhancing its ability to withstand deformation. The higher concentration of crumb rubber has a clear effect on the rheological properties of rubberized bitumen with increased complex modulus, storage modulus, loss modulus, and reduced angle of phase. The addition of crumb rubber in bitumen positively marks the rutting factor, thus, improving the rutting resistance of the rubberized pavement mix. As proved, the high correlation coefficient values are reasonably indicative of an adequate level of consistency on the effect of crumb rubber satisfied on the physical and rheological properties of the rubberized binder.

Davide Lo Presti concentrated on the conduct of the road pavement technology of Rubber Modified Bitumen (RTR-MBs). Indeed, the several assistances to asphalt pavement efficiency and general infrastructure sustainability are so evident that it is highly recommended to consider RTR-MBs techniques as a first alternative to the binders presently used in road pavements. The High Viscosity wet process technology, which has been commonly demonstrated to provide several advantages, enables highway developers in specific to decrease pavement layer thickness owing to rubberized bitumen's demonstrated characteristics.

Rokade's report on the use of LDPE and CRMB indicates that the Marshall Stability value, which is Semi-Dense Bituminous Concrete's strength parameter, showed an increasing trend and the maximum values increased by about 25% by adding LDPE and CRMB. Not only has this research favorably enhanced the waste plastics and tyres in the road construction industry, but it has also effectively enhanced the significant boundaries that will ultimately lead to better and longer highways.

Plastics-waste is covered over aggregate in the altered method (dry process). This enables to bind bitumen better to the plastic-waste covered aggregate owing to enhanced bonding and enhanced contact region between polymer and bitumen. The coating of the polymer also decreases the voids. This prevents trapped air from absorbing moisture and oxidizing bitumen. Weidong Cao (2007) concentrated on the characteristics of recycled tyre rubber modified asphalt mixtures using dry process Testing of three kinds of asphalt mixtures holding distinct carbon content (1%, 2% and 3% by weight of complete blend) and a control combination without rubber.

Based on the outcomes of rutting trials (60 C), indirect tensile tests (10 C) and variance analysis, the inclusion of recycled tire rubber in asphalt mixtures using dry processes could improve the engineering characteristics of asphalt mixtures, and the rubber content has an important impact on the Performance of continuous elevated temperature deformation resistance and low temperature cracking (Table 3).

Table 3. Tests on crumb rubber modified bitumen (crumb-60)

Sr. no.	Test conducted	Test results	Specifications	IS code
1	Specific Gravity	1.03	0.97–1.02	IS-1202
2	Flash point,° C	290	220 °C	IS-1209
3	Thin film oven test			
(A)	Penetration value before conducting TFOT mm	35	30–50	IS-1203
(B)	Reduction in Penetration value after conducting TFOT %	14.28	Max up to 35%	IS-1203
(C)	Loss in mass by heating %	0.075	1	IS-9382
(D)	Increase in softening point, °C, Max	1.00	6 °C	IS-1205
4	Separation Test			
(A)	The softening point before the test, °C	62.00	Min 60 °C	IS-1205
(B)	Softening point after test, °C		The difference in softening point, Max 3 °C	IS-1205
	Top portion	61.00		
	Bottom porion	60.30		

5 Bagasse Ash

Sugarcane Bagasse Ash Sugarcane bagasse is the sugarcane refinement natural product. It is the fiber left to make juices after the process of extraction that comes from the sugar cane. Sugarcane is one of the non-wood fibers species that are heterogeneous, hygroscopic and have a big percentage of very thin-walled cells. When the juice was extracted from sugarcane, the amounts of product volumes in sugar factories make it a waste product. It usually utilizes furnaces as a fuel in the same sugar mill that generates about 8–10% of ashes. Sugarcane bagasse is a natural fiber that is widely used in the construction industry such as banana fiber and softwood pulp. The fiber can be split into two high-E components, asbestos fibers, carbon fiber and glass, and low-E fibers consisting of two components of natural and synthetic fibers. The bagasse is commonly used as a fertilizer and biofuel product because it is crushed in ethanol every 10 tons of sugarcane; almost three tons of moist bagasse is produced by a sugar factory. Because bagasse is a by-product of the cane sugar sector, the manufacturing amount in each nation is consistent with the amount of sugar cane generated.

The high moisture content of bagasse, typically 40–50%, is detrimental to its use as a fuel. In general, bagasse is stored prior to further processing. For electricity production, it is stored under moist conditions, and the mild exothermic process that results from the degradation of residual sugars dries the bagasse pile slightly. For paper and pulp production, it is normally stored wet in order to assist in the removal of the short pith fibers, which impede the papermaking process, as well as to remove any remaining sugar.

M. Zulfikri, M. Zainudin et al. was discovered to be 0.34% ash content focused on Optimum Sugarcane Bagasse. The quantities of SCBA used were 5.63% from 0.1 to 0.5% by weight of the upper limit OAC filler. The study results showed that modified mixtures using SCBA are effective in increasing the Marshall stability, flow, and resilient module by 0.6%, 4.9%, and 17.4% of ordinary HMA respectively. Compared to unmodified mixtures, the VTM is also greater, and this would affect the rutting strength of these mixtures. Air void content of 4.94% for the altered sample is sufficient to provide room for asphalt expansion to prevent bleeding or flushing cabin in order to improve rutting susceptibility and decrease pavement skid resistance. The use of SCBA as a filler in HMA was therefore discovered to be appropriate for use in road pavements and as an alternative material to profit from environmental and financial elements.

Due to the HRS combination with a variety of filler-containing 60%, David D.M. Huwae noted that Bagasse-ash is the ideal composition compared to the others (20%, 40%, 80%, and 100%). The outcome shows that Bagasse-ash is appropriate as a filler based on testing of a variety of Bagasse-ash and cement content (Marshall-stability test 1205,040 kg; flow test 4,427 mm; Marshall-Quotient 273,717 kN/mm; VMA 20,249%; VFA 74,206% and VIM 5,223%). Using Bagasse-ash as a filler can both reduce the need for cement fillers and provide a relatively elevated economic value as well as overcoming current waste.

5.1 Application of Bagasse Ash

5.1.1 Fuel

Bagasse is often used in sugar mills as a primary source of energy. When burned in quantity, it produces enough heat energy to supply with energy to spare all the needs of a typical sugar mill. To this end, cogeneration, the use of a fuel source to supply both heat energy used in the mill and electricity, typically sold on the consumer electrical grid, is a secondary use for this waste product.

5.1.2 Pulp, Paper, Board and Feed

In many tropical and subtropical nations, Bagasse is widely used as a wood replacement for pulp, paper, and board manufacturing such as India, China, Colombia, Iran, Thailand, and Argentina. It generates pulp with physical characteristics that are well suited for generic printing and writing documents as well as tissue products but is also commonly used in the manufacturing of boxes and newspapers. It can also be used to make plywood or particle board-like panels, called bagasse boards, and is regarded as a useful replacement for plywood. It is widely used to make partitions and furniture.

5.1.3 Aggregates

The aggregates used in laboratory research are composed of crushed rock, crushed gravel or other difficult material held on the 2.36 mm sieve. They shall be clean, difficult, and durable, cubic in form, free of dust and soft or friable matter, organic or other harmful matter. 20 mm, 12 mm, 6 mm and dust aggregates used in this research. Tests on aggregates were performed to assess the fundamental physical characteristics along with aggregate gradation (Tables 4, 5, 6).

Bagasse ash

Testing laboratory experiments on both plain and altered binders were performed to evaluate their characteristics and to verify that they fulfill the IS requirements. Tests have shown that the characteristics are within IS boundaries.

Table 4. Chemical composition of Bagasse ash

Item	Percentage (%)
Carbon	10.91
Silica (Sio2)	72.33
Magnesium	0.58
Calcium	0.63
Aluminum (Al2O3)	3.24
Ferrum	0.85

Table 5. Test on aggregates

Sr. no.	Properties tested	Unit	Results	Specification limits for bituminous concrete (BC) as per MoRTH(2001)
1	Los Angeles abrasion value	(Percent)	16.76%	30% (max)
2	Combined flakiness and elongation indices	(Percent)	14.00%	30 (max)
3	Water absorption	(Percent)	1.30%	2 (max)
4	Specific gravity (CA)	(g/c m^3)	2.68	2–3
5	Specific gravity (FA)	(g/c m^3)	2.66	2–3
6	Aggregate impact value	(Percent)	13.60%	24 (max)

Table 6. Blending of aggregates

Proportioning of materials	25 mm	20 mm	12 mm	6.3 mm	Dust	Total
		0.270	0.230	0.190	0.310	1.000

5.2 Marshall Stability Test

The Marshall specimens are ready to grade the DBM grade 2 mixes for the acquired mixture gradation. The samples are ready in accordance with ASTM D6927-06 (Bituminous Mixtures Standard Test for Marshall Stability and Flow). The samples were casted using standard VG 30 binder and Modified Binder Crumb Rubber (CRMB 60). Bagasse ash was used as a partial substitute for the filler to prepare the samples. The bagasse ash used in both standard and crumb rubber modified blend as partial filler substitute was varied from 1% to 3% and the corresponding optimum bitumen content was acquired from Marshall Stability test outcomes. It was noted that 7.5% o can be substituted by bagasse ash, which provides maximum stability and requires marshall values for both simple and altered DBM mixes as per MoRTH.

5.3 Preparation of Specimen

Aggregate fractions of sizes 25 mm, 20 mm, 12 mm, 6 mm and stone-dust (and bagasse ash) passing 2.36 mm and maintained in 75 µ is used as a gradation blend. For both simple and altered bitumen, specimens are cast. The mold's size is 1000 mm in diameter. Aggregates and Binder weight are calculated using Bulk Density from the Marshall Stability Test. For the VG-30 blend, the aggregates are preheated to 165 °C and the CRMB-60 mix to 185 °C. Binder is heated to the consistency of pouring. For the corresponding binder, the aggregates and the binder are blended in the pan and the temperature of the mixture is preserved as per IRC: SP:53. Bagasse ash is weighed while casting the blend with the filler and then added with aggregates to the pan. The blend for the necessary thickness is compacted using UTM well within the compacting temperature. After letting the specimen in the mold for 24 h, the specimen is de-molded and dimensions are measured to ensure the correct thickness.

5.4 Water Sensitivity Test

Marshall Moulds is ready for 7% air voids by compacting the samples on either hand for the comparable water sensitivity test. Indirect tensile strength and the proportion of conditioned to unconditioned specimen are screened for the unconditioned and conditioned samples.

Gradation variation after 10% Bagasse ash is one of the main variables for decreasing strength and keeping the blend temperature becomes hard as the bagasse ash percentage increases.

5.5 Optimum Asphalt Content

Asphalt content varied at 4.00%, 4.25%, 4.5%, 4.75% and 5% by weight of asphalt blend to determine the optimum asphalt content. For each variation in asphalt content, three samples were screened. The optimum asphalt content was determined in accordance with the requirement specification based on the mixed outcomes of the Marshall Test.

5.6 Marshall Stability and Optimum Binder at Varying Crumb Rubber Percentage in VG-30 BC Mix

As the percentage of Crumb rubber replacement increased from 1% to 3%, the stability and optimum binder content of the VG-30 BC mix is increased. Binder content at 3% crumb rubber at an optimum of 5% but greater than the control combination without crumb rubber.

5.7 Marshall Stability and Optimum Binder at Varying Bagasse Ash Percentage in CRMB-60 BC Mix

With the addition of bagasse ash for 10% substitute, stability has risen and after the strength has declined after 10% owing to multiple bagasse ash features. The gradation variation after 10% bagasse ash is one of the main variables for decreasing strength and keeping the mixture temperature becomes hard with the rise in the bagasse ash proportion.

5.8 Effect of Optimum Bagasse Ash on Gradation of Aggregates

The decrease in stability at 15% bagasse ash replacement and further was noticed from Marshall Stability test results, this can be explained by plotting a graph for 10% optimum bagasse ash replacement which clearly indicates the obtained gradation after the replacement is very close to deviating away from the desired mort&h DBM grade 2 gradation.

6 Results and Discussion

The conclusions of this study are defined as the following based on the outcomes of the laboratory investigation Crumb rubber is recommended as an additive in asphalt mixture. Adding a crumb rubber tends to boost asphalt mixture strength and quality. Increased stability and decreased flow are shown. Modified asphalt combination of crumb rubber required less asphalt content. However, the low content of asphalt improves air void in the blend and thus the permeability of blend improves the durability of asphalt blend. Therefore, due to the reduced asphalt content in crumb rubber-modified asphalt blend, there should be more concern about the durability of the asphalt blend. The use of SCBA as a filler in HMA was found suitable to be used for road pavement and as an alternative material that will benefit in environmental and economic aspects.

References

1. Murana, A.A., et al.: Partial replacement of cement with bagasse ash in hot mix asphalt. Niger. J. Technol. (2015). https://doi.org/10.4314/nit.v34i4.5
2. Feroze Rashid, A.M., et al.: Morphological and nanomechanical analyses of ground tire rubber-modified asphalts. In: Innovative Infrastructure Solution. Springer (2016)
3. ASTM standard D 1559-89: Standard test method for resistance to plastic flow of bituminous mixtures using Marshall apparatus 4 inches (100 mm - diameter specimen) intended for mixes containing aggregate up to 1 inch (25.4mm), USA
4. Prasannakumar, C., et al.: Study of Crumb Rubber For the partial Replacement For bitumen or as an Alternative Material for Bitumen. Grenze Scientific Society (2017)
5. Huwae, D.D.M., et al.: Bagasse-ash as filler in HRS (hot rolled sheet) mixture. In: Advances in civil, Environmental, and Materials Research (2016)
6. Presti, D.L.: Recycled tyre rubber modified bitumen for road asphalt mixtures: a literature review (2013). https://doi.org/10.1016/j.conbuildmat.2013.09.007
7. Airey, G., et al.: Manufacturing terminal and field bitumen–tyre rubber blends: the importance of processing conditions (2012). https://doi.org/10.1016/j.sbspro.2012.09.899
8. Silva, H.M.R.D., et al.: Rheological changes in the bitumen caused by heating and interaction with rubber during asphalt–rubber production. Researchgate (2014)
9. IS 73: Indian standard paving bitumen–specification, 2nd revision
10. IRC: 37-2012: Guidelines for the design of flexible pavements. Indian roads congress, New Delhi
11. IRC: SP:53-2010: Guidelines on use of Modified bitumen in Road Construction
12. Kowlaski, K.J., et al.: Eco–Friendly materials for a new concept of asphalt pavement (2016). https://doi.org/10.1016/j.trpro.2016.05.426
13. Lekha, B.M., et al.: Study of addition of waste plastic in dense bituminous macadam with stone dust and bagasse ash as filler. Int. J. Res. Appl. Sci. Eng. Technol. (2016)
14. Ministry of Road Transport & Highways: Specifications for road and bridge works, 5th revision, April 2013
15. Zainudin, M.Z.M., et al.: Effect of sugarcane bagasse ash as filler in hot mix asphalt. Mater. Sci. Forum (2012). https://doi.org/10.4028/www.scientific.net/MSF.846.683
16. Nitish Kumar, K., et al.: Study of using waste rubber tyres in construction of bituminous road. Int. J. Sci. Eng. Res. (2016)
17. Mashaan, N.S., et al.: Effect of crumb rubber concentration on the physical and rheological properties of rubberized bitumen binders. Int. J. Phys. Sci. (2011)
18. Wulandari, P.S., et al.: Use of crumb rubber as an additive in asphalt concrete mixture. Sustain. Civ. Eng. Struct. Constr. Mater. (2017). https://doi.org/10.1016/j.proeng.2017.01.451
19. Khanna, S.K., Justo, C.E.G., Veeraragavan, A.: Highway Engineering, revised 10th addition
20. Khanna, S.K., Justo, C.E.G., Veeraragavan, A.: Highway Material and Pavement Testing
21. Standard Specification for Asphalt-Rubber binder ASTM D 6114-97
22. Ucol-Ganiron Jr., T.: Scrap waste tire as an additive in asphalt pavement for road construction. Int. J. Adv. Appl. Sci. (2013)
23. Cao, W.: Study on properties of recycled tire rubber modified asphalt mixture using the dry process (2007). https://doi.org/10.1016/j.conbuildmat.2006.02.004

General Procedure for Pavement Maintenance/Rehabilitation Decisions Based on Structural and Functional Indices

Hossam S. Abd El-Raof[1], Ragaa T. Abd El-Hakim[1(✉)], Sherif M. El-Badawy[2], and Hafez A. Afify[2]

[1] Faculty of Engineering, Tanta University, Tanta 31527, Egypt
{hussam_saber, ragaa.abdelhakim}@f-eng.tanta.edu.eg
[2] Faculty of Engineering, Mansoura University, Mansoura 35516, Egypt
sbadawy@mans.edu.eg, hafezafify@yahoo.com

Abstract. Due to traffic loading and environmental conditions, pavements deteriorate over time. Other factors that may affect pavement performance are material properties and construction practices. However, it is important for road users to have the road network at a certain acceptable level. Typically, pavement functional indices, such as Pavement Condition Index (PCI), have been conspicuously utilized to determine which type of pavement maintenance/ rehabilitation (M/R) should be applied for a specific pavement of a certain condition. Many researchers concluded that pavement surface condition, in some cases, does not reflect the condition of the underlying layers. Others argued that the treatment decisions based on the functional indices are sometimes overestimated or underestimated. This has galvanized many researchers to consider other indices, among of them are structural indices such as Structural Condition Index (SCI). Many studies recapped that the structural indices lead to more effective M/R decisions. Thus the current research aims to propose a more practical procedure for selecting the most appropriate M/R decision based not only on the functional indices as many highway agencies do but also on the existing structural condition of the degenerated pavement. To develop such procedure, data from 8 Long Term Pavement Performance pavement test sections were evaluated functionally and structurally and the decision was taken based on both functional and structural conditions. The proposed procedure is found to yield reasonable M/R decisions as compared to the use of either one of the indices.

Keywords: Structural condition index · Pavement condition index · M/R decisions · PMS · Overlay design · Maintenance decision tree

1 Introduction

In a Pavement Management System (PMS), flexible pavements can be evaluated functionally or structurally. The functional condition represents the ability of the pavement to carry the future loading at acceptable level of serviceability [1]. The structural condition can be defined as the capability of the pavement to carry the traffic

© Springer Nature Switzerland AG 2020
S. Badawy and D.-H. Chen (Eds.): GeoMEast 2019, SUCI, pp. 13–24, 2020.
https://doi.org/10.1007/978-3-030-34196-1_2

loading over its design period [1]. Pavement distresses and ride quality have been used as the primary indicators for pavement preservation and timing [2]. The most eminent functional indices are Pavement Condition Index (PCI), International Roughness Index (IRI) and Pavement Condition Rating (PCR). These indices assess the current pavement condition contingent on the observable distresses on the pavement surface abjuring the condition of the sublayers. Functional condition is very essential, as it is related to the user's comfort and safety [3]. Nonetheless, these indicators cannot reflect the actual load carrying capacity or the structural condition of the pavement as the comparable structural indicators do [4]. Hence, suggested treatments based on these indicators are often overestimated or under estimated [5]. Chowdhury et al. (2012) stated that, the maintenance activity based on the functional indices is not the optimal treatment [6]. Additionally, Zaghloul et al. (1998) concluded that the results obtained from the functional indices may be independent of the underlying structure [7]. Therefore, the pavement structural condition should be considered for an effective M/R decision [5]. Pavement condition keeps exacerbating, notwithstanding amounts of seal coats and thin overlays are applied by state highway agencies every year. This, in part, is a result of rejecting the structural condition of pavement layers and subgrade soil. Considering the pavement structural condition leads to efficacious treatment decisions [8]. Structural condition is a hidden indicator and is not important to the user [3]. It is not considered in maintenance decision. From engineering view, functional and structural conditions are equally substantial for any PMS. A pavement surface in a poor condition can be construed as in a poor structural condition, however in some cases; a poor surface condition does not mean a poor structural condition [5].

2   Trials to Correlate Structural and Functional Indices

Over the years, many researchers tried to correlate the structural indices with the functional ones. To exemplify, deflection data collected from a section of (I-81) Southbound, Virginia, was utilized to examine the correlation between structural condition of the pavement in terms of Structural Condition Index (SCI) and functional condition of the pavement in terms of Load Related Distress Rating (LDR). The study also tried to collate center deflection and LDR. SCI is the structural condition index and it will be defined in the next section and LDR is a function of the distresses exhibited in the wheel path as presented in Eq. (1) [9]:

$$LDR = Deduct\_Alligator\_Crk - Deduct\_Rutting - Deduct\_Patching \qquad (1)$$

The results indicated that there is no correlation between structural and functional indices [10]. Flora (2009) tried to correlate the Falling Weight Deflectometer (FWD) deflections and functional indices namely IRI, PCR, or the rut depth. The results show that, at 95% confidence level, there is no statistical correlation between these indices [11]. A t-test was conducted between the PCI and Structural Health Index (SHI), which is a structural index developed for Louisiana Department of Transportation, to examine the relation between these two indices statistically. A significant difference was found between the indices represented by a P-value of 0.001. Also, a

Pearson correlation coefficient of 0.41 was found indicating a poor correlation between them [2]. It can be concluded from the previous studies that there are no obvious correlations between structural and functional indices. Based on the previous discussion, it is imperative to realize the importance of conflating both structural and functional indices to obtain a full evaluation for the pavement condition. Thus the main objective of this research is to find a more pragmatic way for an efficacious treatment or rehabilitation decision. That may be achieved by taking the decision based on both structural and functional indices simultaneously. This means, the functional indices can be used to provide an assessment for the current pavement surface condition, and the structural condition would help know the pavement load carrying capacity. In this study the PCI is used to evaluate the pavement condition functionally, and the SCI is considered for the pavement structural evaluation.

3 Structural Condition Index (SCI)

The SCI is defined as the ratio between the SN_{eff} and the required SN (SN_{req}) as presented in Eq. (2) [8].

$$SCI = \frac{SN_{eff}}{SN_{req}} \tag{2}$$

Where
SCI = structural condition index.
SN_{eff} = existing pavement structural number.
SN_{req} = required structural number.

As can be seen, SCI is a simple index and the interpretation of its meaning is straightforward. For SCI value equal to or more than one, the pavement would be intact. It may also be sufficient for the future required Equivalent Single Axel Loads (ESALs). In this case the pavement may only require a preservation maintenance activity (crack sealing or chip sealing). Conversely, the pavement requires a rehabilitation or reconstruction activity if the SCI value is lower than one. The pavement in this case is enfeebled and is not adequate for the estimated future traffic loads. Nam et al. (2016) reformed the decision of M/R based on SCI values as presented in Table 1 [12]. For the prediction of the SN_{eff}, there are many models available in the literature [13, 14]. Based on the comprehensive evaluation of the most known models using the Long Term Pavement Performance (LTPP) data, Abd El-raof et al., 2018b, reported that the calibrated Kavussi et al. model is recommended to calculate SN_{eff} [14]. Calibrated Kavussi et al. model is presented in Eq. (3):

$$SN_{eff} = K_1 * D_0^{K_2} * D_{90}^{K_3} \tag{3}$$

Where:
D_o = peak deflection at a standard 9000-Ib FWD load (microns).
D_{90} = deflection at radial distance of 90 cm from the center of loading plate (microns).
K_1, K_2, and K_3 = regression coefficients = 85.740, −0.770 and 0.310, respectively.

Table 1. Pavement treatment decision based on SCI value (Nam et al. 2016)

SCI *100	M/R decision	Treatment example
>90	Do nothing	—
80–90	PM	Seal Coat Crack seal Thin Overlay (1″-2″)
65–80	LRhb	Seal Crack and place 1.5″ ACP Spot Repair and 1.5″ Overlay Seal Coat and 2″ to 3″ Overlay
50–65	MRhb	Mill ACP and 2″-4″ ACP Overlay Mill 5.5″ ACP and replace with 5.5″ Overlay
<50	HRhb	Remove Existing Pavement, 10″-12″ Lime Treat Subgrade, Place New Flexible Base, and 2″ ACP. Remove 5″ ACP and Place 8″ ACP

On the other hand, the SN$_{req}$ can be estimated according to the expected ESALs accumulated during the desired design period using Eq. (4) [15].

Where PM = Preventative Maintenance, LRhb = Light Rehabilitation, MRhb = Moderate Rehabilitation, HRhb = Heavy Rehabilitation, and ACP = Asphalt Concrete Pavement

$$\log W_{18} = Z_R S_o + 9.36 \log(SN + 1) - 0.2 + \frac{\log[(\Delta PSI)/(4.2 - 1.5)]}{0.4 + 1094/(SN + 1)^{5.19}} + 2.32 \log M_R - 8.07$$

(4)

Where:

W_{18} = 18 Kips (80 KN) equivalent single axle load application number.

Z_R = a normal deviate for a given reliability (R).

S_o = overall standard deviation of traffic.

SN = required structural number (in.).

ΔPSI = loss in serviceability.

MR = subgrade resilient modulus (psi).

4   Pavement Condition Index (PCI)

Based on visual survey, pavement surface condition can be quantified using the traditional PCI. PCI is a numerical value (ranges from 0 to 100) which rates the surface condition of the pavement. The value of 100 represents the best condition, while the worst condition is represented by a value of 0 [16]. It is a tool for rating the pavement and may also be used for the maintenance/rehabilitation alternatives. Continuous observation of the pavement condition and determination of the PCI value can be used to establish pavement deterioration curves which permit the early identification of maintenance and rehabilitation needs. Through PCI, the current design and maintenance procedure can be verified and improved [16]. The PCI value is decreased by a deduct value which depends on the severity and extent of the surface distresses. Severity of each distress can either be Low (L), Medium (M), or High (H). This classification is determined according to the distress level of deterioration. Based on the PCI value, the pavement surface condition can be rated as Good to Failed. In addition, maintenance/rehabilitation decision can be taken according to Table 2.

Table 2. M/R strategy according to PCI value [17]

PCI	Rating	Strategy
85–100	Good	Preventative maintenance
70–85	Satisfactory	Minor rehabilitation
55–70	Fair	Minor rehabilitation
40–55	Poor	Major rehabilitation
25–40	Very poor	Major rehabilitation
10–25	Serious	Reconstruction
0–10	Failed	Reconstruction

5   Data Collection

A total of 8 pavement sections from the Specific Pavement Study (SPS-1) of the Long Term Pavement Program (LTPP) program were used in this research. These sections were selected such that they cover all four climatic regions in the U.S. as well as different subgrade types, traffic levels, and pavement structure layer thicknesses. The climatic regions in the U.S. are classified into wet/freeze, dry/freeze, wet/non freeze, and dry/non freeze and are referred to as WF, DF, WNF, and DNF, respectively [18]. Table 3 presents the structural system, climatic region, subgrade type, and test dates of each section. In addition, the number of FWD data measurements is presented.

Table 3. Main properties of SPS-1 section used for SCI and PCI calculations

Section ID	Climatic region	Layer yhickness, in. (mm)			Subgrade type	Test date
		AC	GB	GS		
01-0101	WNF	7.40 (188)	7.90 (201)	–	A-7-5	4/28/2005
01-0102	WNF	4.20 (107)	12.00 (305)	–	A-7-6	5/28/2002
04-0114	DNF	6.80 (173)	12.00 (305)	–	A-2-4	4/2/2002
10-0101	WF	7.00 (178)	8.10 (206)	39.00 (990)	A-2-4	10/4/2005
10-0102	WF	4.30 (109)	11.80 (300)	39.00 (990)	A-2-4	8/13/1996
19-0101	DF	7.70 (196)	8.00 (203)	25.00 (635)	A-6	2/16/2001
30-0113	DNF	5.80 (147)	8.40 (213)	–	A-1-b	7/16/2001
31-0114	DF	6.60 (168)	12.00 (305)	24.00 (610)	A-7-6	7/10/2000

AC = Asphalt Concrete Layer, GB = Granular Base Layer, GS = Granular Sub-base Layer, DF = Dry, Freeze, DNF = Dry, Non Freeze, WF = Wet, Freeze, and WNF = Wet, Non Freeze.

6   PCI and SCI Calculations

The PCI values were calculated according to the ASTM-D6433 procedure [16]. The PCI values are decreased by the deduct values obtained based on the distress type and its severity. At each test date, the SCI was also calculated for the same sections. SN_{eff} value was calculated using Calibrated Kavussi et al. model as can be shown in Eq. (3), while SN_{req} was calculated using the AASHTO 1993 nomograph. The SCI values for some sections are presented in Fig. 1. In addition, Table 4 gives a recapitulation of SCI and PCI values for the eight used sections.

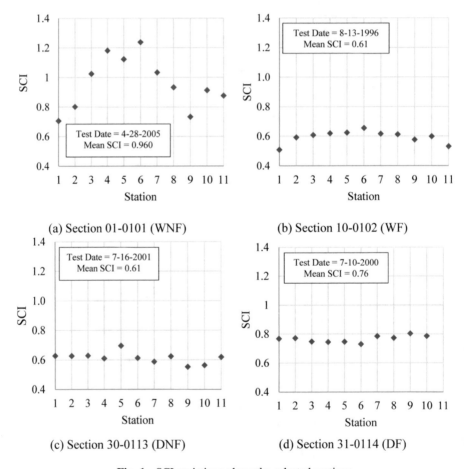

(a) Section 01-0101 (WNF)

(b) Section 10-0102 (WF)

(c) Section 30-0113 (DNF)

(d) Section 31-0114 (DF)

Fig. 1. SCI variations along the selected sections

7 Discussion of Results

It can be seen that the PCI values for 01-0101, 30-0113, and 31-0114 LTPP sections have similar functional condition. However, these sections have disparate SCI values which indicate different structural capacity for future traffic loading. In other words, using the same M/R decision based on the PCI value for these sections would not be fallacious. Thus, the structural condition should be taken into account for more effective maintenance decisions. The same condition exists for Section 01-0102, 04-0114, and 19-0101 with negligible difference in the functional condition as indicated by the PCI values and a significant difference in the structural condition indicated by the SCI values. Moreover, Section 10-0102 and 31-0113 have the same structural condition but they are functionally different. In addition, Section 01-0101 and 04-0114 have the same structural capacity with significantly different PCI values. These results indicate that; the functional indices are independent of the structural condition which confirms the results of other studies such as Zaghloul et al. (1998) [7]. The results also reveal that a high functional performance does not mean a sound structure as shown in the case of Section 10-0102. These results imply that the SCI can discriminate between strong and weak pavements. Therefore, it can be used as a screening tool for network level evaluation. Its simple interpretation and meaning makes it a robust candidate for network level evaluation.

Table 4. Table PCI and SCI values for the selected sections

Section ID	Test date	PCI (%)	SCI
01-0101	4/28/2005	63.00	0.96
01-0102	5/28/2002	31.17	0.58
04-0114	4/2/2002	30.03	1.05
10-0101	10/4/2005	19.08	0.77
10-0102	8/13/1996	85	0.61
19-0101	21/6/2001	38.57	0.80
30-0113	7/16/2001	64.09	0.61
31-0114	7/10/2000	64.46	0.76

8 Selecting the M/R Decision

Based on the PCI and SCI values summarized in Table 4, the appropriate M/R decision can be taken. The threshold values illustrated in Tables 1 and 2 can be used for this purpose. Additionally, the flow chart presented in Fig. 2 is proposed to simplify the decision making. For LTPP Section 01-0101, the SCI value of 0.960 requires "Do Nothing" as it is structurally adequate for the future traffic loading. This means that, the section is able to serve for another design period without any treatment. Conversely, the PCI value of 63% indicates that the pavement condition is fair as presented in Table 2. In other words, light to moderate rehabilitation is recommended based on the PCI value. Thus, the "Do Nothing" treatment based on SCI value is insufficient for the future performance of the section and the user satisfaction. Existence of surface cracks may permit the water to get into the sublayers causing a faster rate of deterioration. In this case preventative treatment such as (Seal Coat or Crack Sealing) will be an effective decision. This proves that, for effective treatment alternative, structural and functional indices should be considered together. Therefore, in Fig. 2 a preventative maintenance is required for a longer future performance than the "Do Nothing" activity. Additionally, a thin overlay can be applied after crack sealing to obtain a smooth surface with low level of roughness and high degree of friction in order to improve the functional performance of the pavement.

For LTPP Section 10-0102, the PCI value of 85% indicates that the pavement surface condition is good. This means a preventative maintenance is required for this section as indicated in Table 2. Conversely, the SCI value of 0.61 implies that this section cannot serve for the desired design period. As indicated in Fig. 2, if a preventative activity is applied without considering the structural condition, this section will significantly deteriorate before the end of the design period. Based on Table 1, this section requires an "MRhb" activity such as "Mill Existing ACP and applying 2″-4″ (5 to 10 cm) ACP overlay". The PCI value indicates that milling the surface layer is uneconomic idea. Additionally, overlaying with 2″-4″ (5 to 10 cm) ACP may be insufficient to carry out the future EASLs. For more effective alternative, both structural and the functional conditions should be considered. Examining the distresses in this section which has a PCI value of 85% indicates little amount of low severity fatigue cracks (9.8 m^2) as reported in LTPP. Instead of milling the whole surface layer area which costs time and money for removing the aged layer and placing the new one, patching the damaged area offers an economic alternative. Patching is the most popular technique of repairing localized areas with intensive cracks. After patching, the required thickness of overlay can be placed without any fear of reflective cracks. The thickness of the overlay can be determined such that the SN_{eff} value becomes $\geq$ 90% from SN_{req}. Equation (5) can be used to estimate the required overlay thickness.

$$d = \frac{SN_{req} - SN_{eff}}{0.44(1 - c)} \tag{5}$$

Where: SN_{eff} = existing structural number. SN_{req} = required structural numberC = coefficient presents the condition of the existing pavement (For flexible pavements C can be assumed (0.5–0.7).

For the remaining sections, the PCI values (19.08 to 64.46) indicate that, pavement surface conditions ranged from serious to fair. On the other hand, SCI values ranged from 0.58 to 1.05 indicates a good structural condition. Selecting the M/R activity based on the low PCI value and ignoring structural conditions leads to an uneconomic decision. For example, Section 10-0101 has a PCI value of 19.08 which indicates a serious pavement condition. In this case the pavement requires a reconstruction strategy to be applied. An SCI value of 0.77 indicate that the pavement is strong and requires a light rehabilitain to be applied. Much time and money will be lost if the reconstruction activity is applied. On the other side, if a light rehabilitation, based on SCI value, is applied reflective cracks may appear after short time from applying it. This discussion

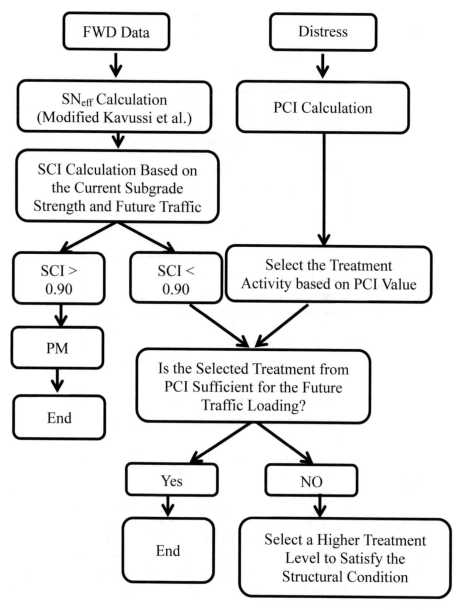

Fig. 2. Flow chart used to select the appropriate M/R based on PCI and SCI

reveals that the structural and functional indicies should be combined for effective M/R decisions. Milling the damaged surface and applying a well designed and constructed overlay layer represnt a robsut alternative in these cases. However, before milling, the existing pavement should be well evaluated. The evaluation of the existing pavement can be conducted based on the flow chart presented in Fig. 3. As can be seen in the figure, FWD testing is required for investigation. The second step is to evaluate the

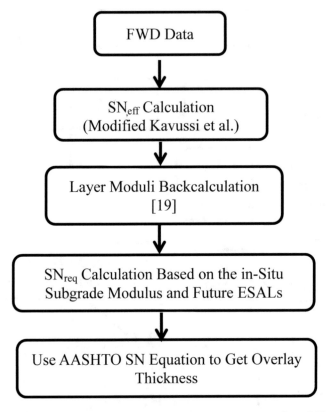

Fig. 3. Suggested procedure for selecting the best M/R decision in case of low PCI and high SCI

existing pavements in terms of SN_{eff} using modified Kavussi et al. model [14]. More investigation is required in terms of AC, base, and subgrade in situ resilient modulus. AC and base layer modulus can be estimated by the recommended procedure by Abd El-raof et al. (2018c) [19], while subgrade resilient modulus can be calculated using Eq. (5). Beside subgrade resilient modulus, future ESALs can be used to calculate SN_{req}. Now the pavement can be milled. With the in situ base modulus, base layer coefficient (a_2) can be calculated and multiplied by the insitu base layer thickness to obtain the contribution of base layer in SN_{req}. AASHTO SN equation can be used to get the accurate overlay thickness after knowing the properties of the mix used in the overlay layer.

$$E_{SG} = -346 + 0.00676 * \left(\frac{2P}{D_{36} + D_{48}} \right) \tag{5}$$

Where:

P = applied load (lbs.). D_{36} and D_{48} = measured deflections at 36 and 48 in. (90 and 122 cm) from the load plate (in.).

9   Summary and Conclusion

Selecting the most appropriate M/R strategy is considered one of the major outcomes of an effective PMS system. The treatment should be selected on the basis of the current condition and the future performance. The existing pavement condition is evaluated by PCI to reflect the actual condition of the surface. On the other hand, SCI is used to reflect the future performance of the pavement under the expected traffic loading. The results indicated that, the functional condition is independent of the structural condition. Additionally, Selecting M/R activity using the structural and functional indices together results in more effective decisions. This tentative study shows the powerful outcome of combining both PCI and SCI in making a credible maintenance/rehabilitation decision.

References

1. Park, K., Thomas, N.E., Wayne Lee, K.: Applicability of the international roughness index as a predictor of asphalt pavement condition. J. Transp. Eng. **133**(12), 706–709 (2007)
2. Elbagalati, O., Elseifi, M., Gaspard, K., Zhang, Z.: Development of the pavement structural health index based on falling weight deflectometer testing. Int. J. Pavement Eng. **19**, 1–8 (2016)
3. Tarefder, R.A., Rahman, M.M., Stormont, J.C.: Alternatives to PCI-based maintenance solutions for airport pavements. In: Transportation Research Board, 94th Annual Meeting, Washington, DC (2015)
4. Rada, G.R., Visintine, B.A., Hicks, R.G., Cheng, D., Van, T.: Emerging tools for use in pavement preservation treatment selection. In: Transportation Research Board, 93rd Annual Meeting, Washington, DC (2014)
5. Bryce, J., Flintsch, G., Katicha, S., Diefenderfer, B.: Developing a network-level structural capacity index for asphalt pavements. J. Transp. Eng. **139**(2), 123–129 (2013)
6. Chowdhury, T., Shekharan, R., Diefenderfer, B.: Implementation of network-level falling weight deflectometer survey. Transp. Res. Record J. Transp. Res. Board **2304**, 3–9 (2012)
7. Zaghloul, S., He, Z., Vitillo, N., Kerr, J.: Project scoping using falling weight deflectometer testing: new jersey experience. Transp. Res. Rec., J. Transp. Res. Board **1643**, 34–43 (1998)
8. Zhang, Z., Claros, G., Manuel, L., Damnjanovic, I.: Evaluation of the pavement structural condition at network level using falling weight deflectometer (FWD) data. In: Transportation Research Board, 82nd Annual Meeting, Washington, DC (2003)
9. Virginia Department of Transportation. State of the Pavement 2006. Richmond (2006)
10. Bryce, J.: A pavement structural capacity index for use in network-level evaluation of asphalt pavements, M.Sc. thesis, Virginia Tech, Blacksburg, VA (2012)
11. Flora, W., Development of a structural index for pavement management: an exploratory analysis, M.Sc. thesis, Purdue University, West Lafayette, IN (2009)
12. Nam, B.H., An, J., Kim, M., Murphy, M.R., Zhang, Z.: Improvements to the structural condition index (SCI) for pavement structural evaluation at network level. Int. J. Pavement Eng. **17**(8), 680–697 (2016)
13. Abd El-Raof, H.S., Abd El-Hakim, R.T., El-Badawy, S.M., Afify, H.A.: Structural number prediction for flexible pavements based on falling weight deflectometer data. In: Transportation Research Board, 97nd Annual Meeting, Washington, DC (2018a)

14. Abd El-Raof, H.S., Abd El-Hakim, R.T., El-Badawy, S.M., Afify, H.A.: Structural number prediction for flexible pavements using the long term pavement performance data. In: IJPE (2018b). https://doi.org/10.1080/10298436.2018.1511786
15. AASHTO, Guide for Design of Pavement Structures, Washington, DC (1993)
16. ASTM, D. Standard Practice for Roads and Parking Lots Pavement Condition Index Surveys, West Conshohocken, Pennsylvania, 6433-07 (2008)
17. Shahin, M.Y., Walter, J.A.: Pavement maintenance management for roads and streets using the PAVER system. Technical Report M-90/05, U.S. Army Corps of Engineers, Washington, DC (1990)
18. LTPP InfoPave (2019). http://www.infopave.com/
19. El-Raof, H.S., El-Hakim, R.T., El-Badawy, S.M., Afify, H.A.: Simplified closed-form procedure for network-level determination of pavement layer moduli from falling weight deflectometer data. J. Transp. Eng., Part B: Pavements **144**(4), 04018052 (2018)

Performance Analysis of Overlays for Flexible Pavement

Ragaa Abd El-Hakim[1]([⊠]), Mustatfa A. K. Abd El-Ghaffar[2],
Mohamed El-Shabrawy M. Ali[2], and Hafez A. Afify[1]

[1] Public Works Department, Faculty of Engineering, Tanta University,
Tanta, Egypt
ragaa.abdelhakim@f-eng.tanta.edu.eg
[2] Public Works Department, Faculty of Engineering, Mansoura University,
Mansoura, Egypt

Abstract. Pavements deteriorate with time in absence of adequate maintenance. The rate of deterioration depends upon a number of factors including
traffic loading (magnitude of wheel load and its repetitions), climate, drainage,
environmental factors and the initial of pavement structure. The design life of
pavement system determines the way in which the surface would deform under
the given traffic and environmental conditions. This in turn would establish the
longitudinal and transverse profile of the riding surface which has to be maintained within acceptable standards, since it influences passenger comfort,
vehicle maintenance, pavement life, etc. In the present study, a number of
parameters were considered in developing a model that predicts the design life
of flexible pavements. These factors were Characteristic Deflection (Dc), Rut
Depth Index (RDI), Cracking Index (CI), Traffic Volume (V), Roughness Index
(RI). For this purpose, 4 road test sections were chosen. These five factors have
been either measured and observed in the field or computed every 3 months
from January 2005 until October 2009. 4 Models were developed for the 4 test
sections and R^2(adj) was between 92.1% and 95.3%.

Keywords: Design life · Dc · RDI · CI · V · EASL · R^2

1 Introduction

The basic need is to introduce a design procedure, handling the variability of deflection
in a consistent manner so as to eliminate the basic risk of localized early failure or over
design initially before strengthen the flexible pavements. Assessment of different
lengths of a pavement sub-sections say 100 m, 150 m, and 200 m can be used to work
out minimum cost solution which takes into account engineering constraints.

S. Badawy and D.-H. Chen (Eds.): GeoMEast 2019, SUCI, pp. 25–41, 2020.
https://doi.org/10.1007/978-3-030-34196-1_3

2   Details of Overlaid Control Road Sections

The field program was established to study the variation of different parameters with time on different type and thicknesses of overlays laid on varied type of existing pavements. Four sections in the region of the Nile Delta have been identified. The details of the identified test sections have been given in Table 1. The above sections have different characteristics in terms of soil subgrades, pavement composition and traffic intensity. On the above sections, continuous measurements were made since 2005. In the present research program, the data has been collected on overlaid flexible pavements and presented chapter 5 for the period of 2005 to 2009.

Table 1. Locational details for road sections under field investigation

Section no.	Road name	Location	Station no.	Sub-sec length
1	Tanta-Quesna	TNT-QSN	6	5 km
2	Benha-Mit Ghamr	BNH-MTG	11	5 km
3	Benha-Quesna	BNH-QSN	111	5 km
4	Kanter Khairia-Bagur	KNK-BGR	115	5 km

3   Field Measurements

For development of suitable methodology, large number of variables such as traffic intensity, pavement composition (four types of subgrade) and road geometrics etc. are required to be incorporated. Four test sections were selected for detailed study during the present study. The final selection was done on the basis of the analysis of deflection data and desirability of having representative test sections for various types of Pavement compositions for long term performance evaluations.

3.1   Deflection Measurements

Structural capacities of flexible pavements can be determined from surface deflection measurements. The most important environmental factor affecting surface deflections of flexible pavements is the temperature of the asphaltic layers (Kim et al. 1995; Shao et al. 1997; Park et al. 2002). All deflection data need to be adjusted to a constant temperature (Chen et al. 2000). BELLS3 and Watson et al. (2004) methods have been used to calculate mid-depth pavement temperature. AASHTO and Chen et al. (2000) approaches have been used to correct pavement deflection to a standard temperature of 30 °C.

The deflection of a pavement surface was measured using the Benkelman beam under vehicle wheels moving at creep speeds (approximately 2 mph). The Benkelman Beam consists of a simple lever arm 3.66 m long supported by 3 legs and pivoted at a distance of 2.44 m from the top (Fig. 1).

The tip of measuring arm was placed between the dual tires of a truck. As the truck moved at the creep speed, the device recorded the rebound deflection of the pavement

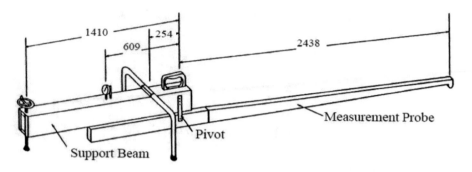

Fig. 1. A Schematic of the Benkelman Beam (all Dimensions in mms).

surface. By suitably placing the probe between the dual wheels of a loaded truck, it is possible to measure the rebound deflection of the pavement structure. While the rebound deflection is related to pavement performance, the residual deflection may be due to non-recoverable deflection of the pavement or because of the influence of the deflection bowl on the front legs of the beam.

The permanent deflection observation points (PDOP) have been established on each road section and marked with paint along 0.90 m from either edge of the lane. Eleven points of the subject road were marked at equal distance of 20 m in each kilometer (to be tested) of the road sub section in both directions, in a homogenous section of about 200 m of the kilometer. Figure 2 illustrates the deflection measurement pavement sub section. Twenty series of deflection measurements were conducted in the present research program. The first series started at January 2005 and the measurements were repeated every three months.

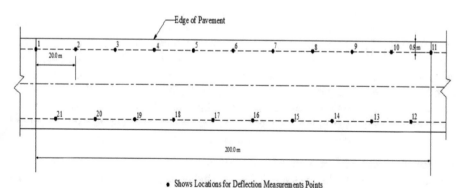

• Shows Locations for Deflection Measurements Points

Fig. 2. Deflection observation points for a sub section length of 200 m at 20 m intervals

3.1.1 Determination of Characteristic Deflection

Overlay design for a given section is based not on individual deflection values but on a statistical analysis of all the measurements in the section, corrected for temperature and seasonal variation. This involves calculation of mean deflection, standard deviation and characteristic deflection as given below.

$$\text{Mean deflection } \bar{x} = \frac{\sum\limits_{i=1}^{n} x_i}{n} \tag{1}$$

$$\text{Standard deviation, } \sigma = \frac{\sqrt{(x - \bar{x}})}{n - 1} \tag{2}$$

Characteristic deflection is given by

$$D_c = \bar{x} + \sigma, \text{ for all other roads}$$

Where, x = individual deflection values (mm)
$\bar{x}$ = mean deflection (mm)
n = number of deflection measurements
σ = standard deviation (mm), and
D_c = characteristic deflection (mm)

Temperature Correction Factor
The stiffness of bituminous layers changes with temperature of the binder and consequently the surface deflections of a given pavement will vary depending upon the temperature of the layers. For the purpose of design, it is necessary that the measured temperature be corrected to a common standard temperature. The standard temperature for Egyptian conditions in plain areas is recommended as 30 °C.

Moisture Correction Factor
Since the pavement deflection is dependent upon the change in the climatic season of the year, it is always desirable to take deflection measurements during the season when the pavement is in its weakest condition (i.e. just after the rainy season) when the subgrade moisture condition is maximum or near saturation. When deflections are measured during other periods of the year, they require a correction factor, which is defined as the ratio of the maximum deflection immediately after rain season to that of the minimum deflection in the dry months. The moisture correction factor is estimated using the values of Plasticity Index and moisture content of the subgrade soil.

Corrected Values of Characteristic Deflection
The ideal time of Benkelman Beam deflection studies is after the rain season. Since the deflection measurements were made in several months of the year, the characteristic deflection was corrected for temperature and moisture as explained above.

3.2   Measurement of Roughness

In order to address specifics of roughness measurement, or issues of accuracy, it is necessary to firstly define the roughness scale. In the interest of encouraging use of a common roughness measure in all significant projects throughout the world, an International Roughness Index (IRI) has been selected. The IRI is so-named because it was a product of the International Road Roughness Experiment (IRRE), conducted by research teams from Brazil, England, France, the United States, and Belgium for the purpose of identifying such an index. The IRRE was held in Brasilia, Brazil in 1982 (Sayers et al. 1986) and involved the controlled measurement of road roughness for a number of roads under a variety of conditions and by a variety of instruments and methods. The roughness scale selected as the IRI was the one that best satisfied the criteria of being time-stable, transportable, and relevant, while also being readily measurable by all practitioners.

Pavement roughness is an important characteristic monitored by many road agencies. It is used as an indicator of road performance and also for feasibility studies. Road roughness is a characteristic of a road surface that is experienced by the operator and passengers of any vehicle travelling over that pavement surface. Surface roughness is a function of the road surface profile and certain parameters of the vehicle including tires, suspension, body mounts, etc. as well as sensibilities of the passenger to acceleration and speed. All of these factors affect the phenomenon of roughness. The road roughness shows the distortion of the pavement surface which contributes to an undesirable or an uncomfortable ride.

3.2.1   Roughness Measurement Methodology

The riding quality (roughness) measurement was carried out using a vehicle mounted bump integrator (VMBI), which records the cumulative vertical displacement of the rear axle relative to the body of the vehicle, while the vehicle is in motion.

The Vehicle Mounted Bump Integrator (VMBI) is a device of the Road Measurement Data Acquisition System (ROMDAS) which is a response-type road roughness meter (RTRRM) mounted in a vehicle to monitor pavement unevenness. It records the vertical displacement of the vehicle chassis relative to the rear axle per unit distance travelled, usually in terms of counts/km or m/km. Since each vehicle responds differently to unevenness due to its own unique springs and shocks, as these changes over time with wear and tear, it is necessary to calibrate each vehicle against a standard unevenness-measuring device. It is also necessary to follow prescribed principles in conducting the measurement to ensure the validity and accuracy of the results. For the present research program, the Roughness was measured using Vehicle Mounted Bump Integrator linked to ROMDAS software controlled by the operator from a laptop computer in the measurement vehicle.

3.3   Rut Depth Measurement

Rutting is a surface depression in the wheel paths caused by inelastic or plastic deformations in any or all of the pavement layers and subgrade. The plastic deformations are typically the result of: (1) densification or one-dimensional compression and consolidation and (2) lateral movements or plastic flow of materials (HMA, aggregate base, and subgrade soils) from wheel loads. The more severe premature distortions or rutting failures are related to lateral flow and/or inadequate shear strength any pavement layer, rather than one-dimensional densification.

Ruts are permanent deformations of the pavement structure. They are an important indicator of the structural integrity of the pavement as well as having an impact on road user safety. A rut is a surface depression in the wheel paths. Pavement uplift may occur along the sides of the rut. However, in many instance ruts are noticeable only after the rains have occurred and the wheel paths are affected by the storage of water. Development of rutting from the permanent deformation in any of the pavement layers or subgrade, usually caused by consolidation or lateral movement of the materials due to traffic loads. Rutting may be caused by elastic movement in the mix in hot weather or inadequate compaction during construction. Significant rutting can lead to major structural failure of flexible pavements.

3.3.1   Procedure

Rut depth in a pavement section is a permanent deformation along the wheel path and it indicates the poor camber profile. The rut depth has been measured in mm by placing a 2.0 m straight edge across the pavement at each PDOP. The above data has been collected as per the procedure shown in Fig. 3 and has been recorded in this chapter.

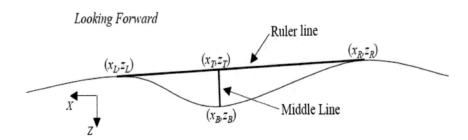

Fig. 3. Rut exteraction description

The rut analysis can be performed in after all the road profile data have been gathered. The rut depth/width is computed according to ASTM 1703 standard (see Fig. 3).

According to this figure, rut depth and rut width are given by:

$$\text{Depth} = \text{sqrt}\left\{ (X_B - X_T)^2 - (Z_B - Z_T)^2 \right\}$$
$$\text{Width} = \text{sqrt}\left\{ (X_R - X_L)^2 - (Z_R - Z_L)^2 \right\}$$

3.4   Measurement of Cracks and Cracking Pattern

Cracking is measured in square meters of surface area. The major difficulty in measuring is that two or three levels of severity often exist within one distressed area. If these portions can be easily distinguished from each other, they should be measured and recorded separately. However, if the different levels of severity cannot be divided easily, the entire area should be rated at the highest severity present. If rutting occurs in the same area, it is recorded separately as its respective severity level.

Cracks are caused by the fatigue failure of the bituminous wearing surface under repeated traffic loading. The cracking initiates at the bottom of the asphalt surface where tensile stress and strength is highest under a wheel load. The cracks propagate on the surface initially as one or more longitudinal parallel cracks. After repeated traffic loading the cracks connect, forming many sided, sharp angled pieces that develop a pattern resembling the skin of an alligator. Alligator cracking occurs only in areas which are subjected to repeated traffic loading. Therefore, it would not occur over an entire area unless it is subjected to traffic loading. Cracking is measured in square meters of surface area.

Typical Cracking Pattern for Traffic Loaded Pavements

3.4.1   Procedure

The crack length have been measured in an area of 1.0 sq.m (1.0 m × 1.0 m) with the PDOP as the center of this square after classifying the cracks into the following three classes:

  i. Fine cracks for width less than 3.0 mm
 ii. Medium cracks of width 3.0 to 6.0 mm
iii. Wide cracks of width greater than 6.0 mm

Cracking is generally regarded as a structural failure, though it greatly affects the riding quality of a pavement which tells us about its functional condition.

3.4.2   Studies for Traffic Volume Counts and Axle Loads

3.4.2.1   Traffic Volume Count Studies

The traffic characteristics are determined in terms of the number of repetitions of an 18.000 Ib (80 kilonewtons (kN)) single-axle loads applied to the pavement on two sets of dual tires. This is usually referred to as the Equivalent Single Axle Load (ESAL). The total ESAL applied on the highway during its design period can be determined only after the design period and the traffic growth factors are known.

The traffic volume data were collected for 24 h. Directional classified volume count was recorded on working day. The volume count stations were fixed by the General Authority for Roads, Bridges and Land Transport (GARBLT) at least 100 m away from an intersection and close to the overlaid test sections.

3.4.2.2   Axle Load Data

As per Asphalt Institute, the traffic characteristics are determined in terms of the number of repetitions of an 18,000 Ib (80 kilonewtons (kN)) single-axle load applied to the pavement on two sets of dual tires. This is usually referred to as the Equivalent Single Axle Load (EASL). The dual tires are represented as two circular plates, each 4.51 in. radius. spaced 13.57 in. apart. This representation corresponds to a contact pressure of 70 Ib/in. The use of an 18,000-Ib axle load is based on the results of experiments that have shown that the effect of any load on the performance of a pavement can be represented in terms of the number of single applications of an 18,000 Ib single axle. Axle load based pavement design methods should be followed for all important roads. The traffic on the roads in developing countries mostly consists of trucks and buses which are normally overloaded. The damaging power of any load with reference to a standard load increment increases approximately by a power of 4. Thus, if the damaging power of an 8 ton single axle truck is 1, the damaging power of a 16 ton single axle load increases to 16 (Kathmandu 1999). By the same way with increasing axle load the surface deflection also will increase.

For highways the maximum legal axle load as specified by the ECP is 8170 kg, with a maximum equivalent single axle load of 4085 kg. The California Bearing Ratio (CBR) method has been modified by the ECP considering the cumulative number of standard axle load repetitions instead of number of commercial vehicles per day. The equivalent single axle loads have been calculated using the following from and from the basic traffic count obtained from the General Authority for Roads, Bridges and Land Transport (GARBLT) traffic count stations on the selected test roads.

3.4.3   Overlay Life Prediction Models

The model has been developed using the field data collected from all the test sections. The following equation is used for developing the general model utilizing the data of test sections located in the Nile Delta.

$$\text{Life of overlay (months)} = a_o + a_1(Dc) + a_2(RDI) + a_3(V) + a_4(CI) + a_5(RI)$$

The characteristic deflection $(Dc = X + \sigma)$ is found out by measuring the deflection at defined points by Benkelman beam deflection method. The mean deflection (X) and the standard deviation (σ) is calculated from the measured deflection at defined points. The details of the commercial vehicles per day [V] have been obtained from the traffic count survey.

The following overlay life prediction models have been developed.

3.4.3.1   General Model to Predict Life of Overlays

(a)   General Model

The following general model has been developed utilizing the data of all the test section located in the region of Nile Delta.

Life of overlay (months) = $-16.0 + 0.233$ Dc $+ 0.586$ RDI $+ 30.9$ ESAL $+ 1.05$ CI $+ 0.00203$ RI

Predictor	Coef.	SE Coef.	T	P
Constant	−16.005	2.483	−6.45	0.000
Dc	0.2333	0.1243	1.88	0.082
RDI	0.5864	0.4158	1.461	0.180
ESAL	30.874	1.667	18.52	0.000
CI	1.0466	0.2926	3.58	0.003
RI	0.002029	0.001233	1.64	0.122

S = 0.0841626 R-Sq = 95.6% R-Sq(adj) = 94.7%

(i)   Model for Benha-Quesna Test section

Life of overlay (months) = $-12.8 + 0.206$ Dc $+ 1.43$ RDI $+ 19.3$ ESAL $+ 0.361$ CI $+ 0.00178$ RI

Predictor	Coef.	SE Coef.	T	P
Constant	−12.7797	0.5646	−22.64	0.000
Dc	0.2062	0.1644	1.25	0.230
RDI	1.4264	0.7776	1.83	0.088
ESAL	19.346	1.046	18.50	0.000
CI	0.3609	0.2190	1.65	0.122
RI	0.0017767	0.0005981	2.97	0.010

S = 0.142319 R-Sq = 96.1% R-Sq(adj) = 95.3%

(ii) Model for Benha-Mit Ghamr Test section

Life of overlay (months) = −25.4 + 0.438 Dc + 0.911 RDI + 29.5 ESAL + 0.902 CI + 0.00627 RI

Predictor	Coef.	SE Coef.	T	P
Constant	−25.393	8.289	−3.06	0.008
Dc	0.4376	0.3253	1.35	0.100
RDI	0.9105	0.4314	2.11	0.053
ESAL	29.521	6.088	4.85	0.000
CI	0.9019	0.4267	2.11	0.053
RI	0.006266	0.003747	1.67	0.117

S = 0.141967 R-Sq = 93.2% R-Sq(adj) = 92.1%

(iii) Model for Tanta-Quesna Test section

Life of overlay (months) = −39.6 + 0.342 Dc + 0.403 RDI + 15.6 ESAL + 0.597 CI + 0.0105 RI

Predictor	Coef.	SE Coef.	T	P
Constant	−39.633	9.587	−4.13	0.001
Dc	0.3423	0.2364	1.45	0.170
RDI	0.4026	0.2341	1.72	0.107
ESAL	15.579	3.715	4.19	0.001
CI	0.5972	0.2525	2.36	0.033
RI	0.010484	0.003732	2.81	0.014

S = 0.116810 R-Sq = 96.8% R-Sq(adj) = 95.3%

(iv) Model for Kanater Khyriya Bagour Test section

Life of overlay (months) = −24.9 + 0.134 Dc + 0.461 RDI + 90.0 ESAL + 0.889 CI + 0.00522 RI

Predictor	Coef.	SE Coef.	T	P
Constant	−24.949	4.432	−5.63	0.000
Dc	0.13404	0.08608	1.56	0.142
RDI	0.4613	0.3464	1.33	0.204
ESAL	90.021	7.033	12.80	0.000
CI	0.88860	0.09996	8.89	0.000
RI	0.005222	0.001685	3.10	0.008

S = 0.0861060 R-Sq = 94.3% R-Sq(adj) = 93.1%

Interpretation of Model to Predict Life of Overlays

As it can be seen through previous Models, the probability of error P is least in most models for equivalent single axle load and that means that the ESALs are the most correlated thing with the age of the overlay, the next is crack index and sometimes roughness index while rut depth index and characteristic deflection is next in correlation with the age of overlay.

From the analysis of variance we can see that we took most of the factors that affect the age of pavement and this is seen through the large value of R-Sq and R-Sq(adj).

It should be reported that for all test sections the program made the analysis of variance which classifies all of the independent variables according to their influence on the dependant variable (age of overlay) and this was characterstic deflection (DC) then rut depth index (RDI), then equivalent single axle load (EASL), then crack index and finally roughness index.

3.4.3.2   General Model to Predict Roughness Index

General Model

Following type of model to predict roughness in terms of characteristic deflection (Dc), rut depth index (RDI), and crack index (CI) has been developed.

$$RI = 1962 + 37.5 \; Dc + 535 \; RDI - 132 \; CI$$

Predictor	Coef.	SE Coef.	T	P
Constant	1961.77	54.24	36.17	0.000
Dc	37.54	52.95	0.71	0.489
RDI	534.73	87.31	6.12	0.000
CI	−131.67	77.04	−1.71	0.107

$S = 36.5121 \; R\text{-}Sq = 92.7\% \; R\text{-}Sq(adj) = 91.8\%$

(i)   **Model for Benha-Quesna Test section**

$$RI = 872 + 121 \; Dc + 705 \; RDI - 144 \; CI$$

Predictor	Coef.	SE Coef.	T	P
Constant	871.78	83.75	10.41	0.000
Dc	121.30	64.84	1.87	0.080
RDI	705.13	83.24	8.47	0.000
CI	−144.14	70.75	−2.04	0.059

$S = 63.0326 \; R\text{-}Sq = 93.7\% \; R\text{-}Sq(adj) = 92.6\%$

(ii)   **Model for Benha-Mit Ghamr Test section**

$$RI = 1963 + 203 \; Dc - 31 \; RDI + 286 \; CI$$

Predictor	Coef.	SE Coef.	T	P
Constant	1962.88	47.11	41.67	0.000
Dc	203.3	106.0	1.92	0.073
RDI	−31.0	155.1	0.20	0.844
CI	285.8	130.8	2.19	0.044

$S = 51.3467 \; R\text{-}Sq = 94.5\% \; R\text{-}Sq(adj) = 93.4\%$

(iii) **Model for Tanta-Quesna Test section**

RI = 2163 + 225 Dc + 167 RDI + 99 CI

Predictor	Coef.	SE Coef.	T	P
Constant	2163.3	137.6	15.72	0.000
Dc	225.5	107.9	2.09	0.053
RDI	167.3	123.8	1.35	0.195
CI	99.5	131.0	0.76	0.459

S = 68.6483 R-Sq = 95.1% R-Sq(adj) = 94.9%

(iv) **Model for Kanater Khyriya Bagour Test section**

RI = 2493 − 11.9 Dc + 350 RDI + 28.9 CI

Predictor	Coef.	SE Coef.	T	P
Constant	2493.24	28.48	87.53	0.000
Dc	−11.90	22.81	−0.52	0.609
RDI	350.44	21.22	16.52	0.000
CI	28.91	16.64	1.74	0.102

S = 23.2210 R-Sq = 96.9% R-Sq(adj) = 95.9%

Interpretation of Model to Predict Roughness Index

As can be observed from previous Models, the probability of error P is least in most models for rut depth index and that means that the RDI is the most correlated distress with the roughness index, and that is practically obvious not only statistically cause when ruts increase the overlay surface will increase in roughness, the next is crack index and sometimes characteristic deflection is next in correlation with roughness.

From the analysis of variance we can see that we took many of the factors that affect the roughness and this is seen through the large value of R-Sq and R-Sq(adj).

And in all test sections the program made the analysis of variance which categorizes all of the independent variables according to their influence on the dependant variable (roughness) and this was characteristic deflection (DC) then rut depth index (RDI), and finally crack index.

3.4.3.3 Model of Correlation Between EASLs and Roughness

General Model

Following type of model to predict roughness in terms of Equivalent Single Axle load has been developed using all the data obtained from field measurements.

RI = 2158 + 1766 ESAL

Predictor	Coef.	SE Coef.	T	P
Constant	2157.79	19.59	110.13	0.000
ESAL	1766.40	19.84	89.02	0.000

S = 35.9972 R-Sq = 94.8% R-Sq(adj) = 94.3%

(i) Model for Benha-Quesna Test section

RI = 1027 + 1513 ESAL

Predictor	Coef.	SE Coef.	T	P
Constant	1027.07	55.58	18.48	0.000
ESAL	1513.09	35.95	42.09	0.000

S = 102.111 R-Sq = 95.0% R-Sq(adj) = 94.9%

(ii) Model for Benha-Mit Ghamr Test section

RI = 2250 + 1909 ESAL

Predictor	Coef.	SE Coef.	T	P
Constant	2249.91	7.43	302.92	0.000
ESAL	1908.77	9.31	205.12	0.000

S = 13.6464 R-Sq = 96.5% R-Sq(adj) = 95.9%

(iii) Model for Tanta-Quesna Test section

RI = 2671 + 1215 ESAL

Predictor	Coef.	SE Coef.	T	P
Constant	2670.99	9.11	293.28	0.000
ESAL	1215.30	7.02	173.18	0.000

S = 16.7333 R-Sq = 96.9% R-Sq(adj) = 96.4%

(iv) Model for Kanater Khyriya Bagour Test section

RI = 2683 + 4995 ESAL

Predictor	Coef.	SE Coef.	T	P
Constant	2683.19	12.26	218.93	0.000
ESAL	4994.95	39.84	125.37	0.000

S = 22.5180 R-Sq = 95.9% R-Sq(adj) = 95.3%

Interpretation of Model to Predict Roughness Index

As can be observed from previous Models, the probability of error P is almost zero in most models and that means that the correlation between roughness index and EASLs is very strong.

3.4.3.4 Estimation of Remaining Service Life Using Prediction Models

Remaining service life (RSL) has been defined as the anticipated number of years that a pavement will be functionally and structurally acceptable with only routine maintenance. The present thesis aimed to use the life prediction models to calculate the remaining service life of a pavement. The following prediction models have been developed in terms of pavement distresses.

General Model

Life of overlay (months) = −56.9 + 0.064 Dc + 4.17 RDI − 3.02 CI + 0.0222 RI

Predictor	Coef.	SE Coef.	T	P
Constant	−56.882	5.550	−10.25	0.000
Dc	0.0636	0.6048	0.11	0.918
RDI	4.171	1.796	2.32	0.035
CI	−3.0244	0.9423	−3.21	0.006
RI	0.02226	0.002812	7.90	0.000

S = 0.410674 R-Sq = 94.3% R-Sq(adj) = 92.9%

(i) Model for Benha-Quesna Test section

Life of overlay (months) = −9.04 − 0.346 Dc + 13.3 RDI − 1.97 CI − 0.00188 RI

Predictor	Coef.	SE Coef.	T	P
Constant	−9.042	2.569	−3.52	0.003
Dc	−0.3462	0.7876	−0.44	0.666
RDI	13.284	2.145	6.19	0.000
CI	−1.9671	0.8735	−2.25	0.040
RI	−0.0019	0.00275	−0.68	0.505

S = 0.693466 R-Sq = 96.9% R-Sq(adj) = 95.8%

(ii) Model for Benha-Mit Ghamr Test section

Life of overlay (months) = −65.0 + 0.707 RDI + 0.365 CI + 0.0241 RI + 0.491 Dc

Predictor	Coef.	SE Coef.	T	P
Constant	−65.039	2.155	−30.17	0.000
Dc	0.4906	0.5141	0.95	0.355
RDI	0.7071	0.6790	1.04	0.314
CI	0.3653	0.6516	0.56	0.583
RI	0.024122	0.001093	22.07	0.000

S = 0.224508 R-Sq = 97.1.0% R-Sq(adj) = 96.7%

(iii) Model for Tanta-Quesna Test section

Life of overlay (months) = $-79.6 - 0.135$ Dc $+ 0.096$ RDI $+ 0.131$ CI $+ 0.0260$ RI

Predictor	Coef.	SE Coef.	T	P
Constant	−79.636	1.378	−57.79	0.000
Dc	−0.1355	0.3005	−0.45	0.659
RDI	0.0961	0.3227	0.30	0.770
CI	0.1315	0.3291	0.40	0.695
RI	0.0260324	0.0006173	42.17	0.000

S = 0.169501 R-Sq = 94.7% R-Sq(adj) = 93.8%

(iv) Model for Kanater Khyriya Bagour Test section

Life of overlay (months) = $-73.3 + 0.283$ Dc $+ 1.63$ RDI $- 0.058$ CI $+ 0.0232$ RI

Predictor	Coef.	SE Coef.	T	P
Constant	−73.331	7.966	−9.21	0.000
Dc	0.2832	0.2937	0.96	0.350
RDI	1.627	1.151	1.41	0.178
CI	−0.0576	0.2316	−0.25	0.807
RI	0.023227	0.003192	7.28	0.000

S = 0.296466 R-Sq = 100.0% R-Sq(adj) = 100.0%

With respect to the developed models to predict life of overlay in months in terms of the pavement distresses, the remaining service life of an existing overlay can be estimated. Since, the constants are known in each developed equation as well as the values of each distress as they are determined from field measurement, the current life or age of overlay in terms of distresses can be determined. Then, if the current life is subtracted from the design life we can get the remaining service life.

In view of the advantages of the adopted tool, it is a simple technique for estimation of the remaining service life of an overlay. Since, it takes into account the existing actual conditions of the pavement which may be named as the structural life of the pavement. In terms of time, the life here means the actual time spent since the first openings of the road for traffic after constructions. If the design assumptions are completely valid, both lives will be compatible. However, in actual practice traffic growth is more than expectations and hence the structural life will be more than the time life. For clarification, if design life of a pavement was 10 years, but deterioration takes place after 5 years i., e before that time. This means that the total design traffic passage used that pavement within 5 years not as expected by the designer within 10 years.

Oppositely, for its limitations it is an empirical model. Also, it assumes that the rate of cracks propagation is constant while this rate should not be so. As the cracks initiated and with more traffic passage over such cracked area, the rate of deterioration is accelerated. Furthermore, the developed models through such technique are based upon only four roads which in turn means low data points for regression. The investigated roads were monitored for five years with regular measurements every 3 months. These models can be easily modified if a data base created through the General Authority for Roads, Bridges and Land Transport (GARBLT),. Once, the actual service life determined, the remaining service life can be easily determined. This can be helpful for optimization the maintenance works in the whole Egyptian roads network.

3.4.3.5   Significance of Variable Distresses in the Life Prediction Model

Each distress has a certain effect on the design life of overlay. In order to find out the most significant distress on the overlay design life, a number of plots have been drawn for each test section representing the relationship between the design life of overlay. These plots were tried with constant values of all distresses except the distress type under consideration.

As can be seen from Figs. 4 and 5 and the statistical model, rutting was found to be the basic parameter affecting design life of overlay. It has the most significant influence as rutting constant in the prediction model is the largest in most of the developed models, so if rutting increases the total life of the pavement increases greatly. The next significant distress is pavement characterstic deflection. It influences the life prediction model in general model, Beha Quesna, Benha Mit Ghamr, and Kanater El-Khayria El-Bagur test sections.

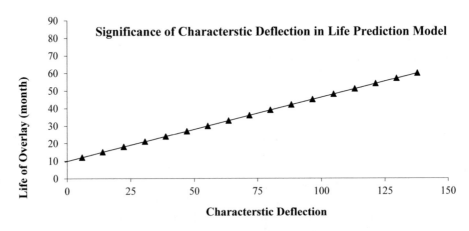

Fig. 4. Significance of Characteristic Deflection in life prediction Model

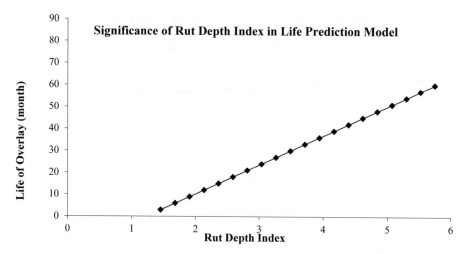

Fig. 5. Significance of Rut Depth Index in life prediction Model

4 Conclusions

1. Rutting was found to be the basic parameter affecting design life of overlay. It has the most significant influence as if rutting increases the total life of the pavement increases.
2. Rutting data can be used to report about pavement condition at network level. It can be used to select the optimum maintenance and rehabilitation option to improve the existing functional characteristics.
3. Remaining service life of an existing pavement can be easily estimated as the actual service life determined by the developed life prediction models.

References

Kim, Y.R., Hibbs, B.O., Lee, Y.C.: Temperature correction of deflections and back-calculated asphalt concrete moduli. Transp. Res. Rec. **1473**, 55–62 (1995)

Shao, L., Park, S.W., Kim, Y.R.: Simplified procedure for prediction of asphalt pavement subsurface temperatures based on heat transfer theories. Transp. Res. Rec. **1568**, 114–123 (1997)

Park, H.M., Kim, Y.R., Park, S.: Temperature correction of multiload-level, falling-weight deflectometer deflections. Transp. Res. Rec. **1806**, 3–8 (2002)

Chen, D.H., Bilyeu, J., Lin, H.H., Murphy, M.: Temperature correction on falling-weight deflectometer measurements. Transp. Res. Rec. **1716**, 30–39 (2000)

Watson, D.E., Zhang, J., Powell, R.B., Murphy, M.: Analysis of temperature data for the national center for asphalt technology test track. Transp. Res. Rec. **1891**, 68–75 (2004)

AASHTO: AASHTO Guide for Design of Pavements Structures. American Association of State Highway and Transportation Officials, Washington, D.C. (1993)

Sayers, M.W., Gillespie, T.D., Queiroz, C.: The International Road Roughness Experiment: Establishing Correlation and a Calibration Standard for Measurements, World Bank Technical Paper No. 45, Washington, D.C. (1986)

Quantifying Effects of Urban Heat Islands: State of the Art

Ragaa Abd El-Hakim[1(✉)] and Sherif El-Badawy[2]

[1] Public Works Engineering Department, Faculty of Engineering,
Tanta University, Tanta, Egypt
ragaa.abdelhakim@f-eng.tanta.edu.eg
[2] Public Works Engineering Department, Faculty of Engineering,
Mansoura University, Mansoura, Egypt
sbadawy@mans.edu.eg

Abstract. Recently, the world has been suffering from the distressing effects of one form of climate change, urban heat island (UHI). It means that urban and suburban areas' air and surface temperatures are hotter than their nearby surrounding rural areas. Pavements and parking lots contributes to about 29% to 45% of the urban areas, and they contribute to the UHI phenomena. During the day, temperature of dark dry surfaces (such as pavements and parking lots) in direct sun can reach up to 88 °C while vegetated surfaces with moist soils might reach only 18 °C under the same conditions. The increase in temperature due to UHI leads to an increase in the peak energy demand using more air conditioners and raising the energy bills. It also leads to an increase in the levels of greenhouse gas emissions (global worming) and air pollution. Increased daytime temperatures, reduced nighttime cooling, and higher air pollution levels related to UHIs affects human health as they lead to general discomfort, respiratory difficulties, heat cramps, and exhaustion. UHI has great and direct effects on the environment, on people and on the human health, on energy consumption and on the economy, and on the pavement performance. The factors that affect the formation and intensity of UHI are versatile in nature. These factors vary between geographic location, time of day and season, synoptic weather, city size, city function and city form. The last two factors are the factors which can be controlled in order to mitigate UHI. Recent studies showed more interest in analyzing and quantifying the UHI phenomenon with more focus on the mitigation techniques. It is abundantly clear that there must be strategies to measure, model and control the phenomenon and achieve one of the Sustainable Development Goals, namely; sustainable cities and communities. The primary focus of this concise, yet comprehensive state of the art paper is to present the different technologies to mitigate the urban heat island. This study presented the different UHI definitions, causes, evaluation methods, and finally compared between the different mitigation techniques and set recommendations and guidelines based on a comprehensive literature review.

Keywords: Albedo · Absorptivity · Emissivity · UHI · UBL · UCL · Cool pavements · UHI mitigation · Cool roofs · Green buildings

© Springer Nature Switzerland AG 2020
S. Badawy and D.-H. Chen (Eds.): GeoMEast 2019, SUCI, pp. 42–69, 2020.
https://doi.org/10.1007/978-3-030-34196-1_4

1 Urban Heat Island-Background

In 1954, 30% of the world's population lived in urban areas, while in 2014, this number increased to 54%, and by 2050, 66% of the world's population is projected to be urban, with a much higher fraction (82%) in the more developed countries [1]. One of the results of such high urbanization is that urban areas exhibit the clearest signs of inadvertent climate modification. The term "heat island" or "reverse oasis" describes the phenomena where urban and suburban areas' air and surface temperatures are hotter than their nearby surrounding rural areas [2–9]. This is depicted in Fig. 1. The first documentation of Urban Heat Island (UHI) was by Luke Howard, 1818, when he found an artificial excess of heat in London city compared to the surrounding Country [10]. The phenomenon is well documented in many metropolitan areas around the world [8, 11–28]. The annual mean air temperature of a city with 1 million people or more can be 1.8–5.4 °F (1–3 °C) warmer than its surroundings. In the evening, the difference can be as high as 22 °F (12 °C) [29]. Generally, the differential warmth of the urban atmosphere is a function of the population of the city [12, 28, 30].

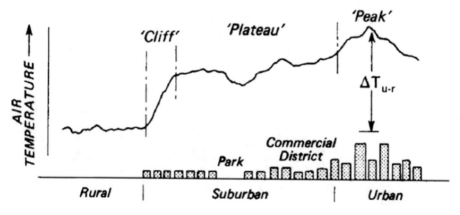

Fig. 1. Generalized cross section of a typical urban heat island [3]

Heat islands are classified into surface heat islands and atmospheric heat islands [31]. They differ in the ways they are formed, the techniques used to identify and measure them, their impacts, and to some degree the mitigation techniques. Table 1 summarizes the basic characteristics of each type of heat island [31].

One has to distinguish between two layers of the atmospheric urban heat island. The first layer termed the **urban canopy layer** (UCL) consisting of the air contained between the urban buildings in the layer of air where people live, from the ground to below the tops of trees and roofs. The second layer, situated directly above the first and called the **urban boundary layer** (UBL) and it starts from the rooftop and treetop level

Table 1. Basic characteristics of surface and atmospheric urban heat islands [31]

Feature	Surface UHI	Atmospheric UHI
Temporal development	• Present at all times of the day and night • Most intense during the day and in the summer	• May be small or non-existent during the day • Most intense at night or predawn and in the winter
Peak intensity (Most intense UHI conditions)	• More spatial and temporal variation: Day: 18 to 27 °F (10 to 15 °C) Night: 9 to 18 °F (5 to 10 °C)	• Less variation: Day: −1.8 to 5.4 °F (−1 to 3 °C) Night: 12.6 to 21.6 °F (7 to 12 °C)
Typical identification method	• Indirect measurement: Remote sensing	• Direct measurement: Fixed weather stations Mobile traverses
Typical depiction	• Thermal image	• Isotherm map • Temperature graph

and extend up to the point where urban landscapes no longer influence the atmosphere [31, 32]. Canopy thermal climate is governed by the immediate site character (building geometry and materials), and not by the accumulation of thermally modified air from upwind areas [33]. All of the urban heat island layers are illustrated in Fig. 2.

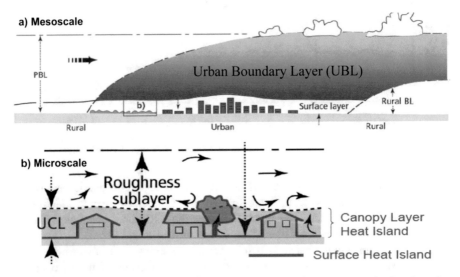

Fig. 2. Schematic representation of surface heat island, the urban canopy and urban boundary layer [34]

2   Urban Heat Island (UHI) Measuring Techniques and Modelling

Urban heat islands may be identified by measuring surface or air temperatures. Surface temperatures have an indirect but significant influence on air temperatures. UHIs have long been studied by ground-based observations taken from fixed thermometer networks or by traverses with thermometers mounted on vehicles. With the advent of thermal remote sensing technology, remote observation of UHIs became possible using satellite and aircraft platforms [32].

Air Temperature Measurements
Air temperature measurements are used to quantify UCL heat island and UBL heat island. Measurement devices could be fixed, traverse, or remote. UCL can be detected by in situ sensors at standard (screen-level) meteorological height or from traverses of vehicle-mounted sensors. The measuring devices could be fixed at weather stations, or traverse using hand-held measurement devices or mounting measurement equipment on cars [32, 34]. Urban Boundary Layer (UBL) heat island observations are made from more specialized sensor platforms such as tall towers, radiosonde or tethered balloon flights, or from aircraft-mounted instruments. These direct in situ measurements require radiation shielding and aspiration to give representative measurements and their setting relative to surrounding features is important [32]. In recent boundary layer studies, remote reading instruments are used to avoid interference with the environment being sensed [3, 35, 36].

Surface Measurements
All surfaces give off thermal energy that is emitted in wavelengths. Instruments on satellites and other forms of remote sensing can identify and measure these wavelengths, providing an indication of temperature. Measurements of surface temperature were used to be done using in situ thermocouple or thermistor thermometry to estimate surface temperatures. An alternative is to use infrared radiometry, where instruments indirectly estimate an apparent surface temperature based upon the radiance received from that area of the surface [37].

Thermal remote sensors observe the surface urban heat island (SUHI), or, more specifically they 'see' the spatial patterns of upwelling thermal radiance received by the remote sensor [32]. Figure 3 illustrates the different measuring techniques for each layer. Figure 4 exemplifies remote sensing images of both Phoenix and Washington DC.

The first SUHI observations (from satellite-based sensors) were reported by [38]. Since then, a variety of sensor-platform combinations (satellite, aircraft, ground based) have been used to make remote observations of the SUHI, or of urban surface temperatures that contribute to SUHI over a range of scales [25, 39–52]. However, surface measurements taken by remote sensing have several limitations. First, they do not fully capture radiant emissions from vertical surfaces, such as a building walls, because the equipment mostly observes emissions from horizontal surfaces such as streets, rooftops, and treetops. Second, remotely sensed data represent radiation that has traveled through the atmosphere twice, as wavelengths travel from the sun to the earth as well as from the earth to the atmosphere. Thus, the data must be corrected to accurately estimate surface properties including solar reflectance and temperature [35].

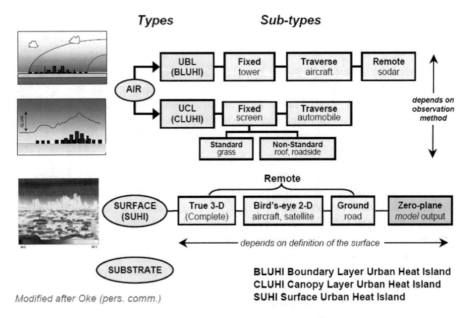

Fig. 3. Urban heat island various measuring techniques [34]

Fig. 4. Remote sensing Image of (1) Phoenix, Arizona, 3 October 2003 11:00 pm to the left, (2) Washington D.C 1 June 2000 to the right via NASA [53, 54]

3   Urban Heat Island-Causes

3.1   Thermal Properties of Urban Materials

In order to understand the causes of Urban Heat Island (UHI) and its mitigation techniques, it is important to study the thermal behavior of urban materials. This requires understanding of the key thermophysical properties of matter that govern thermal phenomenon. These properties can be divided into two distinct categories: (a) heat transfer through a system, (b) thermodynamic or equilibrium state of a system [55]. Heat transfer, can occur by means of radiation, conduction and convection. Heat transfer properties of materials relating to radiation include absorptivity (α_{abs}), albedo (β) and emissivity (ε).

Solar Energy and Absorptivity
About 45% of the solar energy radiates at wavelengths in the visible spectrum, (nominally between 0.3 and 0.7 μm). Also, note that only a little more than 1% of the sun energy at shorter wavelengths (UV and X-solar radiation) and the rest (54%) is in the infrared (IR) region [56]. Radiation energy from sun is either reflected or absorbed. Absorbed energy is then conducted deeper, emitted as radiation, or transferred to air near the surface by convection [57]. Solar energy is a major contributor to the formation of the UHIs. Pavement surfaces will absorb portion of the sun's solar radiation. Thus an increase in the thermal energy occurs. The rate at which radiant energy is absorbed per unit of surface area is dependent on the absorptivity (α_{abs}), *where* $0 \leq \alpha \leq 1$.

Solar Reflectance (Albedo, β)
The rate at which energy is reflected by the surface is known as the albedo (β) of the surface. Albedo is simply $(1 - \alpha_{abs})$ where α_{abs} is the absorptivity of the surface. It takes into account the full spectrum of solar radiation and not just those in the visible range [58, 59]. The albedo of pavement surfaces differs greatly by the materials used in construction. Its values range between zero (complete absorption) and one (complete reflectance). From a climate change perspective, albedo is the first line of defense a surface has against incoming solar radiation. It is regarded as the most important factor in the mitigation of the UHI effect [60]. Reasonable increases in urban albedo can

Table 2. Comparison of the albedo of different surfaces [62–64]

Material type	Albedo
Asphalt	0.05–0.10 (new)- 0.10–0.15 (weathered)
Gray portland cement concrete	0.35–0.40 (new)- 0.20–0.30 (weathered)
White portland cement concrete	0.70–0.80 (new)- 0.40–0.60 (weathered)
Green grace	0.25
Black soil	0.13
Bare soil (Land)	0.17
Desert sand	0.40
Cool pavement coatings	+0.50

achieve a decrease of up to 2 °C to 4 °C in air temperature [61]. Table 2 compares the typical ranges of the albedo for different pavement surfaces [62, 63].

A comparison of the Albedo values of different materials that can be found in the downtown of an urban area is shown in Fig. 5.

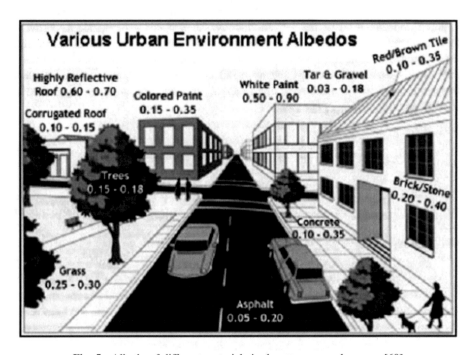

Fig. 5. Albedo of different materials in downtown an urban area [60]

A recent study conducted at Lawrence Berkeley National Laboratory, showed that a 0.1 increase in albedo reduces the pavement temperature by about 4 °C ± 1 °C while a 0.25 increase in the pavement albedo will cause a significant decrease in the pavement temperature by about 10 °C [65].

Emissivity (ε)
Emissivity (ε) is a radiative property of the surface with values in the range $0 \leq \varepsilon \leq 1$. This property provides a measure of how efficiently a surface emits energy relative to a blackbody. It depends strongly on the surface material and finish [55, 65]. Most pavement materials have high emittance values which contributes to the UHI [66].

Representative values of construction materials emissivity as reported in published articles and their validated field measurement values are illustrated in Table 3, [67].

Table 3. Field validation of emissivity [67]

Material	Published emissivity emissivity	Field trial results
Asphalt paving	0.967–0.970	0.95–0.971
Concrete	0.93–0.97	0.90–0.98
Brick	0.93	0.94

The absorptivity and emissivity of a surface are independent of each other as illustrated in Fig. 6 [57].

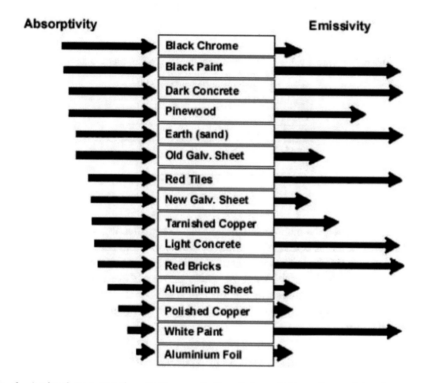

Fig. 6. A visual representation of the magnitudes of absorptivity and emissivity for common building materials [57].

Heat Capacity

Heat capacity or (thermal capacity) specifies the amount of heat energy required to change the temperature of an object by a given amount. It can be calculated as follows [68]:

$$C = \frac{\Delta Q}{\Delta T}$$

Where:

C = heat capacity, in joules per kelvin (J/k)
ΔQ = change in the amount of heat in joules
ΔT = change in temperature in kelvins.

It is more common to report the heat capacity as function of a unit mass such as joules per gram per kelvin (j/g.k). This is called the specific heat capacity (Cp) which is widely used in thermal analysis. Unlike natural materials such as soils and sand, pavements can store more heat. The high heat capacity of the pavement materials contributes to UHIs at night when they resale the stored heat. A comparison of the specific heat capacity values of different materials and natural soil is represented in Table 4 [69].

Table 4. Typical values of specific heat capacity of different materials [69]

Material	Specific heat capacity (j/g.k)
Asphalt	0.92
Concrete	0.88
Granite	0.79
Sand	0.835
Soil	0.80

Thickness

The thickness of a pavement is an important factor affecting how much heat the pavement will store. Of course, thicker pavements store more heat. Thinner pavements will heat faster in the day while cool more quickly at night. A recent study showed that there exists a critical layer thickness at which the maximum surface temperature is optimized (i.e. the maximum surface temperature reaches its lowest point). Pavement thickness greater than this optimum will lead to higher maximum and minimum surface temperatures [70].

3.2   Heat Transfer Models of Pavement Materials: Basics and Principles

Selecting the appropriate paving materials not only ensure stability and safety for road users, but also the ability to mitigate heat absorption and high surface temperatures contributing to the UHI effect and human comfort. In order to select the optimum pavement materials for UHI mitigation, it is also important to understand, study, and develop accurate models for the heat transfer through pavement materials. Literature search has shown that there are several models available [58, 71–79]. In general, heat transfer models predict the surface temperature as well as the temperature profile of the pavement as a function of the climatic data such as (air temperature, wind speed, relative humidity, solar radiation) and material properties such as (albedo, emissivity, aerodynamic roughness, thermal diffusivity, and specific heat of the pavement layer, and thermal diffusivity of the sub-grade and base layers.

3.3   Causes of Urban Heat Island

Several factors contribute to the creation of UHIs. Among these factors reduction of vegetation in urban areas and the solar reflectivity of urban materials are the two primary key factors affecting the formation of UHIs. These two factors are given more attention because unlike other UHI factors, the technology to mitigate them is available via cool roofs and cool pavements. Other factors to consider in the formation of UHIs are heat trapping (heat stored and re-radiated) by urban geometry, the properties of urban surfaces, weather, and geographical location, human-caused (anthropogenic) heat input from transportation and industrial processes [2, 22, 80]. These factors could be classified as controllable and non-controllable factors as illustrated in Fig. 7 [80].

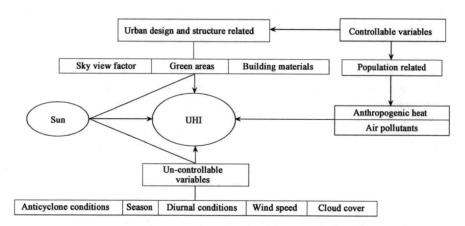

Fig. 7. Generation of urban heat island (UHI) [80].

The UBL can be heated from below as a result of anthropogenic heat input and the formation of UHI in the urban boundary layer (upper part of the UHI) could be due to:

- The slow decline of turbulent sensible heat flux in the late afternoon and evening, and its failure to turn negative at night (unlike rural areas).
- The large release of sensible heat from storage in the urban system in the late afternoon and evening.

There are some features or properties in the Urban Canopy Layer (UCL) and the Urban Boundary Layer (UBL) that could cause positive thermal anomaly due to energy alteration, some of these properties and their consequences are illustrated in Table 5 [81].

Table 5. Suggested "Causes" for the canopy layer and urban boundary layer heat islands (not in rank order) [81]

Altered energy terms giving a positive thermal anomaly	Urban properties underlying energy balance changes
Urban Canopy Layer (UCL)	
1. Increased solar absorption	Canyon geometry - greater surface area and 'trapping' by multiple reflection
2. Increased long-wave radiation from the sky	Air pollution - greater infra-red absorption and reemission
3. Decreased net long-wave radiation loss	Canyon geometry - smaller sky view factor
4. Anthropogenic heat	Buildings and traffic - heat output
5. Increased storage of sensible heat	Building/paving materials and larger surface area - greater thermal admittance
6. Decreased evapotranspiration	Building and paving materials - surface 'waterproofing'
7. Decreased total turbulent heat transport	Canyon geometry - increased wind shelter
Urban Boundary layer (UBL)	
1. Increased absorption of short-wave radiation	Air pollution - increased absorption by aerosol and gases
2. Anthropogenic heat	Chimneys and stacks - heat output
3. Increased sensible heat flux from below	UCL heat island - greater heat flux from roofs and street canyons if warmer than UBL air Rough, warm city - increased entrainment from capping inversion
4. Increased sensible heat from above	

3.4   Modeling of Urban Heat Island Phenomenon in Relation to Pavements

Several attempts have been made to model the phenomenon of Urban Heat Island. Different modeling techniques have been used to model this phenomenon by predicting the asphalt pavement temperature profile. Table 6 shows some examples of these attempts.

4   Urban Heat Island (UHI) Negative Effects

The field of UHI has become highly interesting for scientists and engineers due to its adverse environmental and economic impacts on the society and promising benefits associated with mitigating high heat intensity [80]. Warmer air temperatures due to UHI lead to several adverse effects on people, environment, economy, and even on flexible pavements.

4.1   Negative Effects on the Environment

UHI negative environmental effects in summer are becoming increasingly undebatable, such as the deterioration of the pedestrian-level thermal comfort and the acceleration of

Table 6. Urban heat island modeling techniques

Model type	Output	Case study (Place)	Duration	Inputs	Ref.
3D finite element model	Asphalt pavement temperature	Northeast Portugal	4 months (December 2003 to April 2004)	1. Hourly solar radiation 2. Air temperature 3. Mean daily wind speed	[72]
One dimensional mathematical model	Asphalt pavement temp. profile	Various regions in US&Canada	–	1. Max. air temperature 2. Hourly solar radiation	[75]
One-dimensional mathematical model	Asphalt pavement near surface temperatures	Phoenix, Arizona	3 days	1. Hourly measured solar radiation 2. Air temperature 3. Dew-point temperature 4. Wind velocity data	[58]
Two-dimensional finite-difference model	Asphalt pavement temp. fluctuations	239 stations validated by field data of Alabama	12 month of Data files of 30 Years from 1961 to 1990	1. Climate conditions 2. Thermal and radiative properties of asphalt mixes 3. Surface convective conditions and geometry 4. Solar irradiation	[78]
One-dimensional finite-difference model	Surface temperature, time of wetness, time of freezing events of concrete Pavement	12 representative geographical locations of USA	Typical meteorological one year data files of Weather data from National Renewable Energy Laboratory (NREL)	1. Ambient temperature 2. Dew-point temperature 3. Ambient relative humidity 4. Wind sped 5. Precipitation 6. Cloud cover 7. Incident global horizontal radiation	[82]
One-dimensional model	Pavement temperature of porous asphalt, dense graded asphalt and Portland cement concrete pavements	Phoenix, Arizona	3 days from 12–14 August 2010	1. Pavement thickness 2. Pavement structure 3. Material type 4. Albedo	[83]
One-dimensional mathematical model	Pavement near-surface temperatures	Phoenix, Arizona	Three year from 2004 to 2007	1. Hourly measured solar radiation 2. air temperature 3. Dew-point temperature 4. Wind velocity data	[57]
One dimensional mathematical model	Temperatures of asphalt concrete during summer	Köping, outside Stockholm	Summer 1997 and validated by one day data in July 1998	Hourly values of 1. Solar radiation 2. Air temperature 3. Wind velocity	[73]

photochemical air pollution [84]. The adverse effects of UHI on the environment include the deterioration of living environment [85], elevation in ground-level (tropospheric) ozone concentrations [60, 86–88], increase of pollution levels, and modification of precipitation patterns which may lead to floods [89–91]. Urbanization leads to production of distinctive negative environmental effects, namely CO^2 emission [89, 92–95]. In parallel, several studies have concluded that urban heat island is responsible for higher pollutant concentration over the cities like Tokyo and Paris [96, 97].

Furthermore, UHI has an evident impact on outdoor and indoor thermal comfort. Many studies conclude that higher urban temperatures lower substantially the specific comfort levels [93, 98–106], and the problem is magnified in low income households which suffer from energy poverty [107]. Most of the studies that investigated the impact on the indoor environmental conditions of low income households have shown that indoor temperatures exceeded comfort temperature thresholds while in many cases exceed the maximum allowed indoor temperatures for health reasons [108–120].

4.2   Negative Effects on People and Human Health

UHI has a serious impact on the health conditions of the vulnerable urban population [93]. The World Health Organization, and other national institutions [111–115, 121], recognize that the exposure to high temperatures may cause important cerebrovascular disorders, cardiovascular stress, thermal exhaustion, heart stroke and cardiorespiratory diseases, decrease the viscosity of blood and increase the risk of thrombosis, thermo regulation and impaired kidney function [88, 122–128]. UHI effect exacerbates hot weather events or periods, intensifies the impact of extreme heat events and cause higher fatalities especially in vulnerable populations such as the elderly [129, 130].

There is a strong evidence that during heat waves, hospital admissions and mortality rate related to high ambient temperature increased - in France, Mediterranean, Northern Europe, Spain, Milan, Italy, Budapest, London, Southern Ontario in Canada, Asian cities like Bangkok, Thailand, Delhi, India, urban areas of Bangladesh and Hong Kong [128, 131–143]. According to Centers for Disease Control and Prevention (CDC), excessive heat claims more lives in the United States each year than hurricanes, lightning, tornadoes, floods, and earthquakes combined [144–146]. In Shanghai, as the urban heat island has grown, heat-related mortality rates have increased [147].

Higher urban temperatures are found to have an important impact on mental health [144]. During the heat waves periods in Australia, hospital admissions for mental and behavior disorders increased for ambient temperatures above 27 °C [148]. In parallel, urban warming seems to have an important impact on social behavior increasing the crime rates. It has been demonstrated that, during hot weather and above 32.2 °C, the correlation between the ambient temperature and the probability of collective and assault and domestic crimes are positive and linear [149].

4.3   Negative Effects on Energy Consumption and Economy

UHI has a serious impact on the quality of life of urban citizens. It increases the energy consumption for cooling purposes, and increases the peak electricity demand during the summer period [85, 87, 90, 95, 150–160]. The hourly, daily or monthly electricity consumption increases between 0.5% and 8.5% per degree of temperature rise [161]. The average additional annual cooling energy penalty due to UHI effect was found to be close to 13.1% [159, 162]. Several studies have produced estimates of UHI economic impact, most of these studies conclude that higher ambient temperatures may have a negative relationship to the Gross National Product (GNP) of the countries. However, the results are country specific and vary considerably as a function of the local conditions. For example, the Garnaut Review on Climate change carried out in Australia [163], concluded that a 5 °C temperature increase may result in a reduction of the Australian GNP by 1.3% by 2030. Another study analyzing the relevant data from the last half of the previous century, concluded that higher temperatures reduce economic growth in poor countries but not of the developed ones [164].

4.4 Negative Effects on Flexible Pavement Performance

The increased temperature due to UHI also affects pavement performance. As the air temperature increases, the asphalt is softer at its early ages (usually 3 to 5 years after opening it to traffic) leading to premature permanent deformation (rutting) in the asphalt layer. Over time, as the pavement is exposed to prolonged periods of elevated temperatures it ages and the pavement cracks. A study on the field data of forty-eight street segments paired into 24 high-and low-shade pairs in Modesto, California, U.S. indicated that tree shade was partially responsible for reduced pavement fatigue cracking, rutting, shoving, and other distress. Shaded street was projected to save $7.13/m^2 ($0.66/ft^2) over the 30-year period compared to the unshaded street [165].

Another study investigated how solar radiation affects asphalt pavement performance under the influence of traffic loading by a software simulation model and building a model of pavement with its real physical properties. With a controlled trail, the result showed that pavement stresses are obviously influenced by solar radiations, surface temperature changes, and wheel pressures. Under the same traffic loading, higher surface temperature led to larger stress. This conclusion matched two major reasons of pavement cracking: thermal cracking and fatigue [166].

5 Pavement Share in Urban Areas in and Outside US

Pavements can absorb and store much of the sun's energy contributing to the urban heat island effect. In large urban cities in the U.S., about 29% to 45% of the city's land is covered with pavements [167]. Figure 8 represents the percentage of paved areas in different cities of the United States and the world after [168, 169].

6 Urban Heat Island Mitigation Strategies

The factors that affect the formation and intensity of UHI are versatile in nature. These factors vary between geographic location, time of day and season, synoptic weather, city size, city function and city form. The last two factors are the factors which can be controlled in order to mitigate UHI. Most of the studies investigated the influence of changing city form in terms of materials, geometry or design and green spaces on UHI.

The US Environmental Protection Agency (EPA) calls for three techniques to help mitigate the extreme temperature increases experienced in and around cities:

1. Design and material selection for pavement surfaces (Cool Pavements).
2. Design and material selection for roof structures and surfaces (Cool Roofs).
3. The incorporation of more trees, planting and landscaping elements in urban cities (City Form) [57].

On view of the impacts of UHI, a number of federal, state and local programs aimed at mitigating the UHI effect and its impacts were developed in the 1990s. The Heat Island Reduction Initiative (HIRI), a federal program that includes representatives from NASA, the US Department of Energy, and the US EPA, was initiated in 1997 to mitigate UHI. Since the inception of the project, Lawrence Berkeley National Laboratory (LBNL) has conducted detailed studies to the impact investigate of HIR strategies on heating and cooling energy use of three selected pilot cities [170]. The EPA initiated the implementation of some sustainable practices that would help in mitigating the UHI effect. City planners also should be engaged with global climate scientists to devise contextually relevant strategies to address the urban heat island effect [171]. In this approach, Million Cool Roofs Challenge, a global challenge initiated by a large number of foundations and stake holders to accelerate access to affordable, sustainable cooling through rapid deployment of cool roof materials. The global competition will provide US$2 million in grants to applicants with the most promising ideas and demonstrated success bringing cool roof innovations to scale [172].

Furthermore, due to the important role pavements play in the formation of urban heat island, the Transportation Research Board (TRB) Design and Construction Group has established a "Paving Materials and the Urban Climate" Subcommittee to address the influence of pavements in the formation and mitigation of the UHI and to examine the relationship of pavements to broader climate concerns. The subcommittee's scope includes modeling, design practices, testing, standards development, and planning and policy considerations. The subcommittee inaugurated its activities at the 2008 TRB Annual Meeting [173]. In the following section, the authors will review mitigation strategies and their effects.

6.1   Cool Pavements

Since paved areas represent nearly 30 to 45% of land cover in urban cities, they can absorb and store much of the sun's energy contributing to the urban heat island effect [167]. Some cool pavement efforts were aimed towards reducing the need to pave, particularly over vegetated areas that provide many benefits, including lowering surface and air temperatures. However, this technique is impractical for busy urban areas. Hence, emerged the need for increasing the pavement ability to cool through three different mechanisms to reduce pavement's contribution to the urban heat island: (1) by providing a surface that reflects a greater amount of solar radiation, (2) by increasing

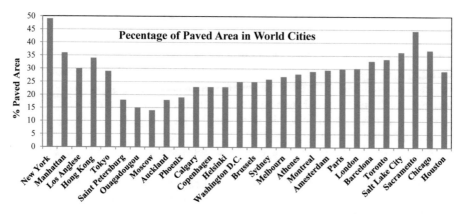

Fig. 8. Percentage of paved area in different world cities

the ability of the pavement to cool at night; and (3) by allowing a pavement to cool through evaporation by designing and building it as a porous structure [174].

Increasing Reflectance of Pavement Surfaces
Although concrete surfaces are already more reflective than asphalt surfaces [175], they can be made even more reflective with the use of white cement and lighter coarse and fine aggregates [59]. However, hence Asphalt pavements account for nearly 94% of pavement in the United States [176], there are several techniques to increase the solar reflectance of asphalt pavements [177] as follows:

1. Surface Gritting with Light-Colored Aggregate
2. Chip Seals with Light-Colored Aggregate
3. Sand Seals with Light-Colored Aggregate
4. Sand- or Shot-Blasting and Abraded Binder Surface
5. Colorless and Reflective Synthetic Binders with Light-Colored Aggregate
6. Surface Painting with Light-Colored Paint
7. Microsurfacing with Light-Colored Materials
8. Grouting of Open-Graded Course with Cementitious Materials.

Increasing the Ability of the Pavement to Cool at Night
A key aspect of urban heat island is the nighttime temperature (the time after sunset and before sunrise). The heat stored in the materials during the day returns back into the air, causing an increase in the air temperature during the night [178]. A study conducted in Arizona studied the influence of pavement materials in lab. And in the field on the near surface pavement temperature and the extent at which the surrounding air temperature will be affected. The materials investigated were; dense graded HMA, porous HMA, conventional Portland Cement Concrete (PCC), and pervious PCC. Based on the observed trends, the section that shows a higher surface temperature during the day tends to have a cooler surface temperature at night, hence less capacity to store heat. Therefore, such materials have less negative contribution to UHI effect [178]. Findings from another study suggested that pervious concrete pavements can provide night time

minimum surface temperatures that are lower than conventional impermeable pavements [179].

Allowing a Pavement to Cool Through Evaporation

Results for analysis of heat gain of a Portland Cement Pervious Concrete (PCPC) system compared with a Portland Cement Concrete (PCC) system showed that stored rain water in the pervious concrete layer had a significant impact on the heat gain in the pervious concrete system, but this rain would then evaporate, improving the heat mitigation by evaporative cooling [180].

A study conducted in Los Angeles (LA) compared between the impact of four mitigation strategies; green roof, cool roof, additional trees, and cool pavement. Comparing the effect of each heat mitigation strategy shows that adoption of additional trees and cool pavements led to the largest spatial-maximum air temperature reductions at 14:00 h (1.0 and 2.0 °C, respectively) [181].

6.2   Cool Roofs

Since 20% of the urban surface is roofed, the widespread use of solar-reflective roofing materials and the use of vegetative – green roofs can save energy, mitigate urban heat islands and slow global warming by cooling roofs [182, 183]. In Australia, the interest to use green roofs is extensively increasing to make buildings more sustainable and provide ecological, and thermal benefits to cities [184]. Many studies combined between different strategies in UHI mitigation [185].

6.3   City Form

Cool city strategies offer energy efficiency improvements [186]. The main strategies related to city form are; cool coatings and vegetation. Although cool roofs are more popular research wise, cool coatings have recently attracted research efforts [187]. Some research results have shown that solutions involving the increase of the global albedo of the city demonstrate the highest benefits, achieving a reduction of peak ambient temperature of up to 3 °C and of peak cooling demand of residential buildings of up to 20% [188]. Three-dimensional greening of buildings is one of the strategies used to change the city form in order to mitigate the effects of UHI. This happens when city planners combine the building roof, wall, balcony, window and other special space with greening design according to the characteristics of different plants [189].

6.4   Case Studies and Regulations in Practice to Control the Phenomenon

Many legislations and regulations in practice are set in different countries to control or mitigate the phenomenon of UHI. At a national level, the Green Building Council of Australia has included green roofs as a 'creditable feature' in the 'Green Star – Design and As Built' rating system, the voluntary sustainability certification for new buildings in Australia [184]. Table 7 shows some of the mitigation techniques and case studies where researchers and city planners have done efforts towards controlling the phenomenon.

7 Summary, Conclusions, and Recommendations

This research extends the knowledge of UHI, and factors affecting its formation, intensity, its adverse effects on the environment, human health, energy consumption, the economy, and on pavement performance. This study also set out to emphasize the UHI measurements and modelling techniques besides mitigation strategies. From the review of the existing body of research about the UHI, the following conclusions can be drawn:

1. UHI causes are versatile, they include geographic location, time of day and season, synoptic weather, city size, city function and city form. However, only two factors affecting UHI formation and intensity are controllable in order to mitigate UHI, namely; city function and city form.
2. There are several forms of measurements and modelling techniques that tried to quantify and describe the UHI phenomenon.
3. The adverse effects of UHI on the environment include: deterioration of living environment, elevation in ground-level (tropospheric) ozone concentrations, raise of pollution levels, modification of precipitation patterns and may lead to floods.
4. It is evident that, when urban heat island grows, heat-related mortality rate increases especially in the vulnerable people like the elderly and the effects of heat waves on human and mental health magnifies.
5. Urban heat island has a serious impact on the quality of life of urban citizens. It increases the energy consumption for cooling purposes, increases the peak electricity demand during the summer period, and can reduce economic growth and GNP of some countries.
6. There is an evident correlation between UHI, and pavement performance. As the air temperature increases the asphalt is softer at its early ages leading to premature rutting in the asphalt layer. Over time, as the pavement is exposed to prolonged periods of elevated temperatures it ages and the pavement cracks.
7. As pavements contribute nearly 29–45% and sometimes up to 49%, one of the most prominent UHI mitigation strategies is cool pavements.
8. Each mitigation technique can contribute to the decrease of elevated temperature. However, combining more than one mitigation strategy helped more to reduce the effects of UHI.

From the comprehensive review of the formation, effects, and mitigation techniques of UHI, the following recommendations could be made:

1. Advanced UHI measuring, quantifying and modelling techniques should be adopted.
2. Combining several UHI mitigation techniques is recommended to enlarge the effectiveness of mitigation.

Table 7. Techniques and case studies of UHI mitigation

Mitigation technique	Temp reduction	Energy saving	Case study	Ref.
Porous asphalt pavement	Nighttime 3.1 °C	N/A	Lab. Arizona	[83]
Pavement thermal properties & albedo	14 °C	Lighting demand 41–57%	Lab and field Arizona	[57]
Unidirectional heat-transfer	6.2 °C (day) and 1.3 °C (night)	N/A	Lab and field China	[190]
Increasing the albedo	2 k	N/A	Simulation time series	[191]
1. Additional trees (Increasing greenery) 2. Cool pavements	1.0 °C at 14:00 2.0 °C at 14:00	N/A	Elmonte, Los Angeles	[181]
1. Water mitigation 2. Cool roofs 3. Cool pavements 4. Increasing greenery	6.0 °C 0.5 °C 2.0 °C 1.5 °C	Cooling demand 50%	Darwin City, Australia	[192]
Combined cool roofs with water technology	1.5 °C, local 10 °C near water technologies	Cooling Demand 39% houses 32% offices	Western Sydney, Australia	[192]
Tree canopy shading plan	Localized 2.5 °C	N/A	Green Square, Australia	[192]
Cool roofs	NM	N/A	LBNL, Berkeley, CA, USA	[182]
Permeable pavements watering	15–35 °C Surface temp.	N/A	Davis, California	[193]
1. Increasing Albedo of roofs 2. Green roofs	1.0.9 K. peak ambient Temp. 2.0.3–3 K Avg. ambient Temp.	N/A	N/A	[183]
Green roofs	0.5 °C at morning 0.3 °C at night	2% per building	Rome	[194]
Green roofs	1.24 °C during Day time	2.5–6%	Constantine, Algeria	[195]
Cool roofs and cool pavements	10 K surface Temp.	17% Cooling Demand	Acharnes Greece	[185]
Small green spaces between buildings	1–4 °C Air Temp.	N/A	Seoul, South Korea	[196]
1. Cool Roofs 2. Green Roofs	Reduce heat gain by 37%, 31%	N/A	Singapore	[197]

References

1. Un, D.: World Urbanization Prospects: The 2014 Revision. United Nations Department of Economics and Social Affairs, Population Division, New York (2015)
2. Gartland, L.M.: Heat Islands: Understanding and Mitigating Heat in Urban Areas. Routledge, London (2012)
3. Oke, T.R.: Boundary Layer Climates. Routledge, London (2002)
4. Asimakopoulos, D., et al.: Energy and climate in the urban built environment. M. Santamouris, University of Athens, Greece (2001). ISBN 1873936907
5. Oke, T.R.: The energetic basis of the urban heat island. Q. J. R. Meteorol. Soc. **108**(455), 1–24 (1982)
6. Oke, T.: The urban energy balance. Prog. Phys. Geogr. **12**(4), 471–508 (1988)
7. Arnfield, A.J.: Two decades of urban climate research: a review of turbulence, exchanges of energy and water, and the urban heat island. Int. J. Climatol. **23**(1), 1–26 (2003)
8. Radhi, H., Sharples, S., Assem, E.: Impact of urban heat islands on the thermal comfort and cooling energy demand of artificial islands—A case study of AMWAJ Islands in Bahrain. Sustain. Cities Soc. **19**, 310–318 (2015)
9. Graves, H., et al.: Cooling Buildings in London: Overcoming the Heat Island. BREPress, Garston (2001)
10. Howard, L.: The Climate of London: Deduced From Meteorological Observations Made in the Metropolis and at Various Places Around It, vol. 2. E. Wilson, London (1833). Harvey and Darton, J. and A. Arch, Longman, Hatchard, S. Highley [and] R. Hunter
11. Tan, J., et al.: The urban heat island and its impact on heat waves and human health in Shanghai. Int. J. Biometeorol. **54**(1), 75–84 (2010)
12. Oke, T.R.: City size and the urban heat island. Atmos. Environ. (1967) **7**(8), 769–779 (1973)
13. Katsoulis, B., Theoharatos, G.: Indications of the urban heat island in Athens, Greece. J. Clim. Appl. Meteorol. **24**(12), 1296–1302 (1985)
14. Balling Jr., R.C., Cerveny, R.S.: Long-term associations between wind speeds and the urban heat island of Phoenix, Arizona. J. Clim. Appl. Meteorol. **26**(6), 712–716 (1987)
15. Lee, D.O.: Urban warming?—an analysis of recent trends in London's heat island. Weather **47**(2), 50–56 (1992)
16. Saitoh, T., Shimada, T., Hoshi, H.: Modeling and simulation of the Tokyo urban heat island. Atmos. Environ. **30**(20), 3431–3442 (1996)
17. Yamashita, S.: Detailed structure of heat island phenomena from moving observations from electric tram-cars in metropolitan Tokyo. Atmos. Environ. **30**(3), 429–435 (1996)
18. Kim, Y.-H., Baik, J.-J.: Maximum urban heat island intensity in Seoul. J. Appl. Meteorol. **41**(6), 651–659 (2002)
19. Figuerola, P.I., Mazzeo, N.A.: Urban-rural temperature differences in Buenos Aires. Int. J. Climatol. **18**(15), 1709–1723 (1998)
20. Kłysik, K., Fortuniak, K.: Temporal and spatial characteristics of the urban heat island of Łódź, Poland. Atmos. Environ. **33**(24), 3885–3895 (1999)
21. Wilby, R.L.: Past and projected trends in London's urban heat island. Weather **58**(7), 251–260 (2003)
22. Jin, H., Cui, P., Huang, M.: Investigation of urban microclimate parameters of city square in Harbin. In: Mediterranean Green Buildings & Renewable Energy, pp. 949–963. Springer (2017)
23. Aflaki, A., et al.: Urban heat island mitigation strategies: A state-of-the-art review on Kuala Lumpur, Singapore and Hong Kong. Cities (2016)

24. Ichinose, T., Shimodozono, K., Hanaki, K.: Impact of anthropogenic heat on urban climate in Tokyo. Atmos. Environ. **33**(24), 3897–3909 (1999)
25. Streutker, D.R.: A remote sensing study of the urban heat island of Houston, Texas. Int. J. Remote Sens. **23**(13), 2595–2608 (2002)
26. Solecki, W.D., et al.: Mitigation of the heat island effect in urban New Jersey. Glob. Environ. Chang. Part B Environ. Hazards **6**(1), 39–49 (2005)
27. Elsayed, I.S.: Mitigation of the urban heat island of the city of Kuala Lumpur, Malaysia. Middle East J. Sci. Res. **11**(11), 1602–1613 (2012)
28. Oke, T., Hannell, F.: The form of the urban heat island in Hamilton, Canada, vol. 108. WMO Technical Note (1970)
29. Oke, T.: Urban climates and global environmental change. In: Thompson, R.D., Perry, A. (eds.) Applied Climatology: Principles & Practices, pp. 273–287. Routledge, New York (1997)
30. Park, H.-S.: Features of the heat island in Seoul and its surrounding cities. Atmos. Environ. (1967) **20**(10), 1859–1866 (1986)
31. EPA: Compendium of Strategies Urban Heat Island Basics. Reducing Urban Heat Islands (2009). https://www.epa.gov/heat-islands/heat-island-compendium
32. Voogt, J.A., Oke, T.R.: Thermal remote sensing of urban climates. Remote Sens. Environ. **86**(3), 370–384 (2003)
33. Oke, T.R.: The distinction between canopy and boundary-layer urban heat islands. Atmosphere **14**(4), 268–277 (1976)
34. Voogt, J.: How researchers measure urban heat islands. In: United States Environmental Protection Agency (EPA), State and Local Climate and Energy Program, Heat Island Effect, Urban Heat Island Webcasts and Conference Calls (2007)
35. EPA: Measuring Heat Island. United States Environmental Protection Agency. https://www.epa.gov/heat-islands/measuring-heat-islands. Accessed 1 Jan 2017
36. Dwivedi, A., Khire, M.: Measurement technologies for urban heat islands. Int. J. Emerg. Technol. Adv. Eng. **4**(10), 539–545 (2014)
37. Voogt, J.A., Oke, T.R.: Complete urban surface temperatures. J. Appl. Meteorol. **36**, 1117–1131 (2011)
38. Rao, P.: Remote sensing of urban heat islands from an environmental satellite. Amer Meteorological Soc 45 Beacon St, Boston, MA 02108–3693, p. 647 (1972)
39. Shen, H., et al.: Long-term and fine-scale satellite monitoring of the urban heat island effect by the fusion of multi-temporal and multi-sensor remote sensed data: a 26-year case study of the city of Wuhan in China. Remote Sens. Environ. **172**, 109–125 (2016)
40. Li, X., et al.: Remote sensing of the surface urban heat island and land architecture in Phoenix, Arizona: combined effects of land composition and configuration and cadastral–demographic–economic factors. Remote Sens. Environ. **174**, 233–243 (2016)
41. Coutts, A.M., et al.: Thermal infrared remote sensing of urban heat: hotspots, vegetation, and an assessment of techniques for use in urban planning. Remote Sens. Environ. **186**, 637–651 (2016)
42. Rotem-Mindali, O., et al.: The role of local land-use on the urban heat island effect of Tel Aviv as assessed from satellite remote sensing. Appl. Geogr. **56**, 145–153 (2015)
43. Hu, L., Brunsell, N.A.: A new perspective to assess the urban heat island through remotely sensed atmospheric profiles. Remote Sens. Environ. **158**, 393–406 (2015)
44. Wu, H., et al.: Assessing the effects of land use spatial structure on urban heat islands using HJ-1B remote sensing imagery in Wuhan, China. Int. J. Appl. Earth Obs. Geoinform. **32**, 67–78 (2014)
45. Imhoff, M.L., et al.: Remote sensing of the urban heat island effect across biomes in the continental USA. Remote Sens. Environ. **114**(3), 504–513 (2010)

46. Tran, H., et al.: Assessment with satellite data of the urban heat island effects in Asian mega cities. Int. J. Appl. Earth Obs. Geoinform. **8**(1), 34–48 (2006)
47. Streutker, D.R.: Satellite-measured growth of the urban heat island of Houston, Texas. Remote Sens. Environ. **85**(3), 282–289 (2003)
48. Voogt, J.A.: Image representations of complete urban surface temperatures. Geocarto Int. **15**(3), 21–32 (2000)
49. Voogt, J.A., Oke, T.: Effects of urban surface geometry on remotely-sensed surface temperature. Int. J. Remote Sens. **19**(5), 895–920 (1998)
50. Nichol, J.: Visualisation of urban surface temperatures derived from satellite images. Int. J. Remote Sens. **19**(9), 1639–1649 (1998)
51. Lo, C.P., Quattrochi, D.A., Luvall, J.C.: Application of high-resolution thermal infrared remote sensing and GIS to assess the urban heat island effect. Int. J. Remote Sens. **18**(2), 287–304 (1997)
52. Roth, M., Oke, T., Emery, W.: Satellite-derived urban heat islands from three coastal cities and the utilization of such data in urban climatology. Int. J. Remote Sens. **10**(11), 1699–1720 (1989)
53. Kaloush, K.E.: Climate change impacts on pavement engineering. In: International Sustainable Pavements Workshop: Airlie Center, Warrenton, Virginia (2010)
54. Washington, DC. Nasa Earth Observatory, Images (2000). Accessed 8 July 2019. https://earthobservatory.nasa.gov/images/928/washington-dc
55. Bergman, T.L., et al.: Fundamentals of Heat and Mass Transfer. Wiley, Hoboken (2011)
56. Geyer, M., Stine, W.B.: Power from the Sun (Powerfromthesun. net). JT Lyle Center (2001)
57. Kaloush, K.E., et al.: The thermal and radiative characteristics of concrete pavements in mitigating urban heat island effects (2008)
58. Gui, J., et al.: Impact of pavement thermophysical properties on surface temperatures. J. Mater. Civ. Eng. **19**(8), 683–690 (2007)
59. American Concrete Pavement Association: Albedo: a measure of pavement surface reflectance. Concr. Pavement Res. Technol. **3**(05), 1–2 (2002)
60. Rosenfeld, A.H., et al.: Cool communities: strategies for heat island mitigation and smog reduction. Energy Build. **28**(1), 51–62 (1998)
61. Taha, H.: Urban climates and heat islands: albedo, evapotranspiration, and anthropogenic heat. Energy Build. **25**(2), 99–103 (1997)
62. Levine, K.: Cool pavements research and technology (2011)
63. Pomerantz, M., et al.: The effects of pavements' temperatures on air temperatures in large cities (2000)
64. Nichols (Nichols Consulting Engineers, C., CTL Group, Cool Pavements Study (Final), Prepared for: City of Chula Vista (2012)
65. Lin, J.D., et al.: The study of pavement surface temperature behavior of different permeable pavement materials during summer time. In: Advanced Materials Research. Trans Tech Publ. (2013)
66. Thermal emittance, in wikipedia (2016). https://en.wikipedia.org/wiki/Thermal_emittance
67. Golden, J.S., Kaloush, K.E.: Mesoscale and microscale evaluation of surface pavement impacts on the urban heat island effects. Int. J. Pavement Eng. **7**(1), 37–52 (2006)
68. Goss, D.J., Petrucci, R.H.: General Chemistry Principles & Modern Applications, Petrucci, Harwood, Herring, Madura: Study Guide. Pearson/Prentice Hall, Upper Saddle River (2007)
69. Mosca, G., Ruskell, T., Tipler, P.A.: Physics for Scientists and Engineers Study Guide, vol. 1. Macmillan, New York (2003)

70. Gui, J., et al.: Impact of pavement thickness on surface diurnal temperatures. J. Green Build. **2**(2), 121–130 (2007)
71. Herb, W., et al.: Simulation and characterization of asphalt pavement temperatures. Road Mater. Pavement Des. **10**(1), 233–247 (2009)
72. Minhoto, M., et al.: Predicting asphalt pavement temperature with a three-dimensional finite element method. Transp. Res. Rec. J. Transp. Res. Board **1919**, 96–110 (2005)
73. Hermansson, Å.: Simulation model for calculating pavement temperatures including maximum temperature. Transp. Res. Rec. J. Transp. Res. Board **1699**, 134–141 (2000)
74. Qin, Y., Hiller, J.E.: Modeling temperature distribution in rigid pavement slabs: impact of air temperature. Constr. Build. Mater. **25**(9), 3753–3761 (2011)
75. Solaimanian, M., Kennedy, T.W.: Predicting maximum pavement surface temperature using maximum air temperature and hourly solar radiation. Transp. Res. Rec. (1417) (1993)
76. Bentz, D.: A Computer Model to Predict the Surface Temperature and Time-of-Wetness of Concrete Pavements and Bridge Decks. National Institute of Standards and Technology, US Department of Commerce (2000)
77. Ramadhan, R.H., Wahhab, H.I.A.-A.: Temperature variation of flexible and rigid pavements in Eastern Saudi Arabia. Build. Environ. **32**(4), 367–373 (1997)
78. Yavuzturk, C., Ksaibati, K., Chiasson, A.: Assessment of temperature fluctuations in asphalt pavements due to thermal environmental conditions using a two-dimensional, transient finite-difference approach. J. Mater. Civ. Eng. **17**(4), 465–475 (2005)
79. Hermansson, Å.: Mathematical model for paved surface summer and winter temperature: comparison of calculated and measured temperatures. Cold Reg. Sci. Technol. **40**(1), 1–17 (2004)
80. Rizwan, A.M., Dennis, L.Y., Chunho, L.: A review on the generation, determination and mitigation of Urban Heat Island. J. Environ. Sci. **20**(1), 120–128 (2008)
81. Oke, T.: The heat island of the urban boundary layer: characteristics, causes and effects. In: Wind Climate in Cities, pp. 81–107. Springer (1995)
82. Bentz, D.P.: A computer model to predict the surface temperature and time-of-wetness of concrete pavements and bridge decks. US Department of Commerce, Technology Administration, National Institute of Standards and Technology (2000)
83. Stempihar, J., et al.: Porous asphalt pavement temperature effects for urban heat island analysis. Transp. Res. Rec. J. Transp. Res. Board **2293**, 123–130 (2012)
84. Taha, H., Konopacki, S., Akbari, H.: Impacts of lowered urban air temperatures on precursor emission and ozone air quality. J. Air Waste Manag. Assoc. **48**(9), 860–865 (1998)
85. Konopacki, S., Akbari, H.: Energy savings for heat-island reduction strategies in Chicago and Houston (including updates for Baton Rouge, Sacramento, and Salt Lake City). Lawrence Berkeley National Laboratory (2002)
86. Lai, L.-W., Cheng, W.-L.: Air quality influenced by urban heat island coupled with synoptic weather patterns. Sci. Total Environ. **407**(8), 2724–2733 (2009)
87. Stathopoulou, E., et al.: On the impact of temperature on tropospheric ozone concentration levels in urban environments. J. Earth Syst. Sci. **117**(3), 227–236 (2008)
88. Kleerekoper, L., van Esch, M., Salcedo, T.B.: How to make a city climate-proof, addressing the urban heat island effect. Resour. Conserv. Recycl. **64**, 30–38 (2012)
89. Zhang, X.Q.: The trends, promises and challenges of urbanisation in the world. Habitat Int. **54**, 241–252 (2016)
90. Yuan, F., Bauer, M.E.: Comparison of impervious surface area and normalized difference vegetation index as indicators of surface urban heat island effects in Landsat imagery. Remote Sens. Environ. **106**(3), 375–386 (2007)

91. O'Malley, C., et al.: Urban Heat Island (UHI) mitigating strategies: a case-based comparative analysis. Sustain. Cities Soc. **19**, 222–235 (2015)
92. Wang, Y., Chen, L., Kubota, J.: The relationship between urbanization, energy use and carbon emissions: evidence from a panel of Association of Southeast Asian Nations (ASEAN) countries. J. Clean. Prod. **112**, 1368–1374 (2016)
93. Santamouris, M.: Regulating the damaged thermostat of the cities—status, impacts and mitigation challenges. Energy Build. **91**, 43–56 (2015)
94. Al-mulali, U., Sab, C.N.B.C., Fereidouni, H.G.: Exploring the bi-directional long run relationship between urbanization, energy consumption, and carbon dioxide emission. Energy **46**(1), 156–167 (2012)
95. Santamouris, M., Paraponiaris, K., Mihalakakou, G.: Estimating the ecological footprint of the heat island effect over Athens. Greece. Climatic Change **80**(3–4), 265–276 (2007)
96. Sarrat, C., et al.: Impact of urban heat island on regional atmospheric pollution. Atmos. Environ. **40**(10), 1743–1758 (2006)
97. Yoshikado, H., Tsuchida, M.: High levels of winter air pollution under the influence of the urban heat island along the shore of Tokyo Bay. J. Appl. Meteorol. **35**(10), 1804–1813 (1996)
98. Bartzokas, A., et al.: Climatic characteristics of summer human thermal discomfort in Athens and its connection to atmospheric circulation. Nat. Hazards Earth Syst. Sci. **13**(12), 3271–3279 (2013)
99. Krüger, E., et al.: Urban heat island and differences in outdoor comfort levels in Glasgow, UK. Theor. Appl. Climatol. **112**(1–2), 127–141 (2013)
100. Orosa, J.A., et al.: Effect of climate change on outdoor thermal comfort in humid climates. J. Environ. Health Sci. Eng. **12**(1), 1 (2014)
101. Thorsson, S., et al.: Potential changes in outdoor thermal comfort conditions in Gothenburg, Sweden due to climate change: the influence of urban geometry. Int. J. Climatol. **31**(2), 324–335 (2011)
102. Hedquist, B.C., Brazel, A.J.: Seasonal variability of temperatures and outdoor human comfort in Phoenix, Arizona, USA. Build. Environ. **72**, 377–388 (2014)
103. Papanastasiou, D., Melas, D., Kambezidis, H.: Air quality and thermal comfort levels under extreme hot weather. Atmos. Res. **152**, 4–13 (2015)
104. Giannopoulou, K., et al.: The influence of air temperature and humidity on human thermal comfort over the greater Athens area. Sustain. Cities Soc. **10**, 184–194 (2014)
105. Katavoutas, G., Georgiou, G.K., Asimakopoulos, D.N.: Studying the urban thermal environment under a human-biometeorological point of view: the case of a large coastal metropolitan city, Athens. Atmos. Res. **152**, 82–92 (2015)
106. Cheung, C.S.C., Hart, M.A.: Climate change and thermal comfort in Hong Kong. Int. J. Biometeorol. **58**(2), 137–148 (2014)
107. Kolokotsa, D., Santamouris, M.: Review of the indoor environmental quality and energy consumption studies for low income households in Europe. Sci. Total Environ. **536**, 316–330 (2015)
108. Sakka, A., et al.: On the thermal performance of low income housing during heat waves. Energy Build. **49**, 69–77 (2012)
109. Wright, A., Young, A., Natarajan, S.: Dwelling temperatures and comfort during the August 2003 heat wave. Build. Serv. Eng. Res. Technol. **26**(4), 285–300 (2005)
110. Lomas, K.J., Kane, T.: Summertime temperatures and thermal comfort in UK homes. Build. Res. Inf. **41**(3), 259–280 (2013)
111. Organization, W.H.: Large analysis and review of European housing and health status (LARES). WHO Regional Office for Europe, Copenhagen (2007)

112. Zavadskas, E., Raslanas, S., Kaklauskas, A.: The selection of effective retrofit scenarios for panel houses in urban neighborhoods based on expected energy savings and increase in market value: The Vilnius case. Energy Build. **40**(4), 573–587 (2008)

113. Summerfield, A., et al.: Milton Keynes Energy Park Revisited: changes in internal temperatures. In: Proceedings of Comfort and Energy Use in Buildings: Getting them Right, NCEUB International Conference. Citeseer (2006)

114. Wingfield, J., et al.: Evaluating the impact of an enhanced energy performance standard on load-bearing masonry domestic construction: understanding the gap between designed and real performance: lessons from Stamford Brook (2011)

115. Mavrogianni, A., et al.: London housing and climate change: impact on comfort and health-preliminary results of a summer overheating study. Open House Int. **35**(2), 49 (2010)

116. Pantavou, K., et al.: Evaluating thermal comfort conditions and health responses during an extremely hot summer in Athens. Build. Environ. **46**(2), 339–344 (2011)

117. Gobakis, K., et al.: Development of a model for urban heat island prediction using neural network techniques. Sustain. Cities Soc. **1**(2), 104–115 (2011)

118. Mihalakakou, G., et al.: Simulation of the urban heat island phenomenon in Mediterranean climates. Pure. appl. Geophys. **161**(2), 429–451 (2004)

119. Mihalakakou, G., et al.: Application of neural networks to the simulation of the heat island over Athens, Greece, using synoptic types as a predictor. J. Appl. Meteorol. **41**(5), 519–527 (2002)

120. Livada, I., et al.: Determination of places in the great Athens area where the heat island effect is observed. Theor. Appl. Climatol. **71**(3–4), 219–230 (2002)

121. Britain, G.: English House Condition Survey 1991: Energy Report. Great Britain, Department of the Environment (1996)

122. Filleul, L., et al.: The relation between temperature, ozone, and mortality in nine French cities during the heat wave of 2003. Environ. Health Perspect. **114**, 1344–1347 (2006)

123. Flynn, A., McGreevy, C., Mulkerrin, E.: Why do older patients die in a heatwave? QJM **98**(3), 227–229 (2005)

124. Hajat, S., et al.: Impact of high temperatures on mortality: is there an added heat wave effect? Epidemiology **17**(6), 632–638 (2006)

125. Kovats, R.S., Kristie, L.E.: Heatwaves and public health in Europe. Eur. J. Public Health **16**(6), 592–599 (2006)

126. Ledrans, M., et al.: Heat wave 2003 in France: risk factors for death for elderly living at home. Epidemiology **15**(4), S125 (2004)

127. Linares, C., Diaz, J.: Impact of high temperatures on hospital admissions: comparative analysis with previous studies about mortality (Madrid). Eur. J. Public Health **18**(3), 317–322 (2008)

128. Rydin, Y., et al.: Shaping cities for health: complexity and the planning of urban environments in the 21st century. Lancet **379**(9831), 2079 (2012)

129. Rosenfeld, A.H., et al.: Mitigation of urban heat islands: materials, utility programs, updates. Energy Build. **22**(3), 255–265 (1995)

130. Patz, J.A., et al.: Impact of regional climate change on human health. Nature **438**(7066), 310–317 (2005)

131. Baccini, M., et al.: Heat effects on mortality in 15 European cities. Epidemiology **19**(5), 711–719 (2008)

132. Chan, E.Y.Y., et al.: A study of intracity variation of temperature-related mortality and socioeconomic status among the Chinese population in Hong Kong. J. Epidemiol. Community Health **66**(4), 322–327 (2012)

133. Changnon, S.A., Kunkel, K.E., Reinke, B.C.: Impacts and responses to the 1995 heat wave: a call to action. Bull. Am. Meteorol. Soc. **77**(7), 1497–1506 (1996)

134. Diaz, J., et al.: Effects of extremely hot days on people older than 65 years in Seville (Spain) from 1986 to 1997. Int. J. Biometeorol. **46**(3), 145–149 (2002)
135. Dousset, B., et al.: Satellite monitoring of summer heat waves in the Paris metropolitan area. Int. J. Climatol. **31**(2), 313–323 (2011)
136. Goggins, W.B., et al.: Effect modification of the association between short-term meteorological factors and mortality by urban heat islands in Hong Kong. PLoS ONE **7** (6), e38551 (2012)
137. Huynen, M.M., et al.: The impact of heat waves and cold spells on mortality rates in the Dutch population. Environ. Health Perspect. **109**(5), 463 (2001)
138. Kovats, R.S., Hajat, S., Wilkinson, P.: Contrasting patterns of mortality and hospital admissions during hot weather and heat waves in Greater London, UK. Occup. Environ. Med. **61**(11), 893–898 (2004)
139. Loughnan, M.E., Nicholls, N., Tapper, N.J.: The effects of summer temperature, age and socioeconomic circumstance on Acute Myocardial Infarction admissions in Melbourne, Australia. Int. J. Health Geogr. **9**(1), 1 (2010)
140. Pirard, P., et al.: Summary of the mortality impact assessment of the 2003 heat wave in France. Euro Surveill. Bull. Eur. Sur Mal. Transm. Eur. Commun. Dis. Bull. **10**(7), 153–156 (2005)
141. Smoyer-Tomic, K.E., Kuhn, R., Hudson, A.: Heat wave hazards: an overview of heat wave impacts in Canada. Nat. Hazards **28**(2–3), 465–486 (2003)
142. Wilkinson, P., et al.: Cold comfort: the social and environmental determinants of excess winter death in England, 1986–1996 (2001)
143. Dhainaut, J.-F., et al.: Unprecedented heat-related deaths during the 2003 heat wave in Paris: consequences on emergency departments. Crit. Care **8**(1), 1 (2003)
144. Number of Heat-Related Deaths: Center for Disease Control and Prevention (2012). https://www.cdc.gov/mmwr/preview/mmwrhtml/mm6136a6.htm. Accessed 5 Jan 2017
145. Climate Change Indicators in the United States: Heat-Related Deaths. EPA, United States Environmental Protection Agency, August 2016. Accessed 5 Jan 2017. https://www.epa.gov/climate-indicators/climate-change-indicators-heat-related-deaths
146. Klinenberg, E.: Heat Wave: A Social Autopsy of Disaster in Chicago. University of Chicago Press, Chicago (2015)
147. Stone, B., Hess, J.J., Frumkin, H.: Urban form and extreme heat events: are sprawling cities more vulnerable to climate change than compact cities. Environ. Health Perspect. **118**(10), 1425–1428 (2010)
148. Hansen, A., et al.: The effect of heatwaves on mental health in a temperate Australian city. Epidemiology **19**(6), S85 (2008)
149. Cohn, E.G.: Weather and crime. Br. J. Criminol. **30**(1), 51–64 (1990)
150. Cartalis, C., et al.: Modifications in energy demand in urban areas as a result of climate changes: an assessment for the southeast Mediterranean region. Energy Convers. Manag. **42**(14), 1647–1656 (2001)
151. Davies, M., Steadman, P., Oreszczyn, T.: Strategies for the modification of the urban climate and the consequent impact on building energy use. Energy Policy **36**(12), 4548–4551 (2008)
152. Dhalluin, A., Bozonnet, E.: Urban heat islands and sensitive building design–A study in some French cities' context. Sustain. Cities Soc. **19**, 292–299 (2015)
153. Fung, W., et al.: Impact of urban temperature on energy consumption of Hong Kong. Energy **31**(14), 2623–2637 (2006)
154. Hassid, S., et al.: The effect of the Athens heat island on air conditioning load. Energy Build. **32**(2), 131–141 (2000)

155. Hirano, Y.: The effects of urban heat island phenomenon on residential and commercial energy consumption. Environ. Syst. Res. **26**, 527–532 (1998)
156. Hirano, Y., Fujita, T.: Evaluation of the impact of the urban heat island on residential and commercial energy consumption in Tokyo. Energy **37**(1), 371–383 (2012)
157. Kikegawa, Y., et al.: Impacts of city-block-scale countermeasures against urban heat-island phenomena upon a building's energy-consumption for air-conditioning. Appl. Energy **83** (6), 649–668 (2006)
158. Kolokotroni, M., Giannitsaris, I., Watkins, R.: The effect of the London urban heat island on building summer cooling demand and night ventilation strategies. Sol. Energy **80**(4), 383–392 (2006)
159. Santamouris, M., et al.: On the impact of urban climate on the energy consumption of buildings. Sol. Energy **70**(3), 201–216 (2001)
160. Taha, H., et al.: Residential cooling loads and the urban heat island—the effects of albedo. Build. Environ. **23**(4), 271–283 (1988)
161. Santamouris, M., et al.: On the impact of urban heat island and global warming on the power demand and electricity consumption of buildings—A review. Energy Build. **98**, 119–124 (2015)
162. Kolokotroni, M., et al.: London's urban heat island: Impact on current and future energy consumption in office buildings. Energy Build. **47**, 302–311 (2012)
163. Garnaut, R.: Garnaut Climate Change Review–Update 2011 Update Paper four: Transforming rural land use (2011)
164. Dell, M., Jones, B.F., Olken, B.A.: Climate change and economic growth: evidence from the last half century, National Bureau of Economic Research (2008)
165. McPherson, E.G., Muchnick, J.: Effects of street tree shade on asphalt concrete pavement performance (2005)
166. Zhang, K.: The effect of urban heat islands and traffic wheel pressure on the performance of asphalt pavements. 2015 NCUR (2015)
167. Ferguson, B., et al.: Reducing urban heat islands: compendium of strategies-cool pavements (2008)
168. Akbari, H., Menon, S., Rosenfeld, A.: Global cooling: Effect of urban albedo on global temperature. Lawrence Berkeley National Laboratory (2008)
169. Scruggs, G.: How Much Public Space Does a City Need? Inspiring Better Cities (2015)
170. Konopacki, S., Akbari, H.: Energy savings for heat-island reduction strategies in Chicago and Houston (including updates for Baton Rouge, Sacramento, and Salt Lake City) (2002)
171. Corburn, J.: Cities, climate change and urban heat island mitigation: localising global environmental science. Urban Stud. **46**(2), 413–427 (2009)
172. Bender, N.: Global Million Cool Roofs Challenge. https://www.k-cep.org/insights/news/million-cool-roofs-launch/. Accessed 25 June 2019
173. Kaloush, K.: Paving materials and the urban climate. In: TR News, p. 11 (2007)
174. Systematics, C.: Cool pavement report, EPA cool pavements study—task 5 (2005)
175. Chen, J., et al.: Field and laboratory measurement of albedo and heat transfer for pavement materials. Constr. Build. Mater. **202**, 46–57 (2019)
176. Pavement Facts, Washington Asphalt Pavement Association (2014). http://www.asphaltwa.com/welcome-facts/. Accessed 23 June 2019
177. Tran, N., et al.: Strategies for design and construction of high-reflectance asphalt pavements. Transp. Res. Rec. **2098**(1), 124–130 (2009)
178. Pourshams-Manzouri, T.: Pavement temperature effects on overall urban heat island. Arizona State University (2013)
179. Carlson, J., et al.: Evaluation of in situ Temperatures, Water Infiltration and Regional 1 Feasibility of Pervious Concrete Pavements 2 3 (2008)

180. Haselbach, L., et al.: Cyclic heat island impacts on traditional versus pervious concrete pavement systems. Transp. Res. Rec. **2240**(1), 107–115 (2011)
181. Taleghani, M., et al.: The impact of heat mitigation strategies on the energy balance of a neighborhood in Los Angeles. Sol. Energy **177**, 604–611 (2019)
182. Levinson, R., et al.: A novel technique for the production of cool colored concrete tile and asphalt shingle roofing products. Sol. Energy Mater. Sol. Cells **94**(6), 946–954 (2010)
183. Santamouris, M.: Cooling the cities–a review of reflective and green roof mitigation technologies to fight heat island and improve comfort in urban environments. Sol. Energy **103**, 682–703 (2014)
184. Pianella, A., et al.: Green roofs in Australia: Review of thermal performance and associated policy development (2016)
185. Kolokotsa, D.D., et al.: Cool roofs and cool pavements application in Acharnes, Greece. Sustain. Cities Soc. **37**, 466–474 (2018)
186. Shickman, K., Rogers, M.: Capturing the true value of trees, cool roofs, and other urban heat island mitigation strategies for utilities. In: Energy Efficiency, pp. 1–12 (2019)
187. Pisello, A.L.: State of the art on the development of cool coatings for buildings and cities. Sol. Energy **144**, 660–680 (2017)
188. Santamouris, M., et al.: On the energy impact of urban heat island in Sydney: climate and energy potential of mitigation technologies. Energy Build. **166**, 154–164 (2018)
189. Cui, Y.-Q., Zheng, H.-C.: Impact of three-dimensional greening of buildings in cold regions in China on urban cooling effect. Procedia Eng. **169**, 297–302 (2016)
190. ShengYue, W., et al.: Unidirectional heat-transfer asphalt pavement for mitigating the urban heat island effect. J. Mater. Civ. Eng. **26**(5), 812–821 (2013)
191. Akbari, H., Matthews, H.D.: Global cooling updates: reflective roofs and pavements. Energy Build. **55**, 2–6 (2012)
192. Santamouris, M., Ding, L., Osmond, P.: Urban heat island mitigation. In: Decarbonising the Built Environment, pp. 337–355. Springer (2019)
193. Li, H., et al.: The use of reflective and permeable pavements as a potential practice for heat island mitigation and stormwater management. Environ. Res. Lett. **8**(1), 015023 (2013)
194. Battista, G., et al.: Green roof effects in a case study of Rome (Italy). Energy Procedia **101**, 1058–1063 (2016)
195. Sahnoune, S., Benhassine, N.: Quantifying the impact of green-roofs on urban heat island mitigation. Int. J. Environ. Sci. Dev. **8**(2), 116 (2017)
196. Park, J., et al.: The influence of small green space type and structure at the street level on urban heat island mitigation. Urban For. Urban Green. **21**, 203–212 (2017)
197. Yang, J., et al.: Green and cool roofs' urban heat island mitigation potential in tropical climate. Sol. Energy **173**, 597–609 (2018)

Use of Fourier Transform Infrared (FT-IR) Spectroscopy to Determine the Type and Quantity of Rejuvenator Used in Asphalt Binder

L. Noor[1], N. M. Wasiuddin[1(✉)], Louay N. Mohammad[2], and D. Salomon[3]

[1] Louisiana Tech University, 600 W. Arizona Ave., P.O.BOX 10348 Ruston, LA 71272, USA
{lno003,wasi}@latech.edu
[2] Department of Civil and Environmental Engineering, Louisiana State University, 3240S Patrick F. Taylor Hall/LTRC, Baton Rouge, LA 70803, USA
louaym@lsu.edu
[3] Pavement Preservation Systems, LLC, 8100 W. Marigold St., Garden City, USA
delmar@technopave.com

Abstract. Recent development in chemical characterization of asphalt binder ascends the scope of using nondestructive techniques to evaluate asphalt binder more rapidly in different laboratory and field condition. Among these techniques, portable Fourier Transform Infrared (FT-IR) Spectroscopy is the most advantageous considering its applicability and efficacy in terms of quick identification and quantification of chemical components in asphalt binder. Previous studies with FT-IR primarily focused on envisaging the chemical alteration in asphalt binder due to the addition of rejuvenator. In this study, FT-IR was used as a primary tool to identify the type and amount of rejuvenator added in asphalt binder. Two types of rejuvenators (bio and aromatic) were used in unmodified, polymer modified and extracted binder from RAP (Reclaimed Asphalt Pavement). Results showed that bio rejuvenator added two distinctive functional groups in asphalt binder at wavenumber 1744 cm^{-1} (C=O stretching) and 1162 cm^{-1} (C-O stretching). The amount of bio rejuvenator in asphalt sample can also be determined since C=O and C-O functional groups increased linearly with the increased percentage of added rejuvenator. The R^2 values observed for unmodified and modified binders are greater than 0.95 for C=O and C-O respectively, indicating the viability of using FT-IR spectrometer for determining the type and amount of rejuvenator added. It is also observed that bio rejuvenator can be determined in samples containing 20% RAP binder mixed with both unmodified and polymer modified binder while achieving a R^2 value of 0.96 (C=O and C-O). On the other hand, aromatic rejuvenator used in this study, did not add any significant functional group in all cases.

© Springer Nature Switzerland AG 2020
S. Badawy and D.-H. Chen (Eds.): GeoMEast 2019, SUCI, pp. 70–84, 2020.
https://doi.org/10.1007/978-3-030-34196-1_5

1 Introduction

Fourier Transform Infrared (FT-IR) Spectroscopy is a rapid and nondestructive technique to evaluate the surface property of a material. It has already been used in identification of asphalt binder source, modifiers such as Styrene-Butadiene-Styrene (SBS) and Polyolefins (PE), crumb rubber, warm mix additives, polyphosphoric acid (PPA). In addition, aging mechanism were also investigated by FT-IR [1, 2].

In FT-IR, an infrared spectra was collected by passing an infrared light through the sample. Absorption occurs when the vibrational frequency of a bond is equivalent to the frequency of the IR. After that, Fourier Transform is performed on the collected signal for further analysis. Based on the material state (liquid, solid or gas), different sampling interfaces such as - ATR (attenuated total reflectance), diffuse reflectance, external reflectance can be used to analyze the chemical property of the material. Among this, ATR with diamond crystal is suitable for material with high absorbance capacity. Sample thickness and amount of sample does not interfere the spectra quality in ATR-FTIR which makes the sample preparation technique easier. Again, the absorbance intensity in ATR-FTIR spectra is linearly proportional to the sample concentration which is useful for quantitative analysis of the substance [3].

In previous studies, FT-IR has been used as supporting technique with other chemical and rheological tests to evaluate the rejuvenation mechanism in asphalt binder. Both petroleum and plant-based rejuvenators were investigated to envisage the effect of the rejuvenation process. Among these studies, waste cooking oil was mostly studied as a petroleum-based rejuvenator. Studies have shown that waste engine oil acted as a rejuvenator and can restore the properties of aged binder [4, 5]. On the other hand, sources of the bio rejuvenator were highly distributed from waste cooking oil to plant derived oils. Waste cooking oil showed improvement in restoring the properties of the aged asphalt binder after proper chemical modification [6–10]. Researchers have shown that aging mechanism in asphalt binder was reversed by soybean oil due to the changes in carbonyl (C=O) and sulfoxide (S=O) functional groups in FT-IR spectra during the rejuvenation process [11]. In addition, 10% bio rejuvenator can restore the aging mechanism of PAV (Pressure Aging Vessel) aged binder [12]. Studies also showed that 5% bio-based rejuvenators (seed oil, cashew nutshell oil and tall oil) can exhibit aging restoration mechanism by unstiffened the RAP binder. It was concluded that bio rejuvenator added a functional group form easter at wavenumber 1750 cm^{-1} and increased C=O functional group in FT-IR spectra [13]. Bio rejuvenator obtained from sawdust can be used at a 15% dosage for the rejuvenation of lower graded PAV (Pressure Aging Vessel) aged binder. Besides, it was also concluded that S=O and C=C indices reflected the rejuvenation process more precisely than C=O index which was increased due to the C=O groups incorporated form the bio rejuvenator [14]. Previous studies also presented the rejuvenation mechanism in unmodified and SBS (Styrene-Butadiene-Styrene) modified asphalt binder and concluded that rejuvenation process was unable to restore SBS crosslinking structures. It showed that physical properties of rejuvenated SBS modified asphalt binder were more stable than the rejuvenated unmodified asphalt [15]. Recent studies also re-recommended the necessity of characterizing rejuvenated binder based on their chemical properties to understand the interaction between asphalt binder and rejuvenator, [16] as well as addressing the recent challenges on implementation of the rejuvenating agent in field application and to understand the chemical characterization of rejuvenation agent and their interaction on polymer modified binders [17].

Previous literatures mainly focused on evaluating the restoration mechanism of rejuvenator due to aging and quantifying the optimum dosage of rejuvenator. However, there is a research gap to quantify the amount of rejuvenator present in asphalt binder. This paper addresses this issue by suggesting an approach of using FT-IR to quantify the amount of rejuvenator present in asphalt binder. In this study, this quantification process was conducted by analyzing absorbance value, absorbance height and absorbance area of the representative functional group added in the asphalt binder due to rejuvenator. Neat binder, polymer (SBS) modified binder and RAP induced modified binder were included to observe the variation in the corresponding functional group responsible for rejuvenator in asphalt binder. Both bio and aromatic rejuvenator were added to evaluate their effect on asphalt binder.

The remaining segments of this study are consisting of experimental methodology with the description of materials, preparation of samples, sampling technique in FT-IR and spectral analysis methods. This segment was followed by results and discussion in which findings from the analysis methods were presented. Finally, in conclusion, segments findings of this study were outlined.

2   Methodology

2.1   Materials

In this study, PG58-28 binder was used as a base binder. Commercially available plant-based rejuvenator was used as a bio rejuvenator, whereas petroleum extracted residual oil was used as an aromatic rejuvenator. Properties of the rejuvenators used in this study are listed in Table 1. In this study, Styrene-Butadiene-Styrene (SBS) polymer was used for the modification of the base binder in which styrene-butadiene ratio was 31/69. The polymer structure was radial and was provided in powder form from the manufacturer. RAP (Recycled Asphalt Pavement) aggregate was collected from the construction plant stockpile. After collection, proper sieving was conducted to find the nominal size of RAP aggregate to carry out the extraction of RAP binder.

Table 1. Material properties of bio and aromatic rejuvenators

Property	Bio-Rejuvenator	Aromatic rejuvenator
Appearance, 77 °F (25 °C)	Dark-Colored Liquid	Dark-Colored Liquid
Odor	Mild Hydrocarbon	Slight
Density, 77 °F (25 °C)	7.6 lb/gal	8.0 lb/gal
Viscosity	100 cP; 77 °F (25 °C)	4.29 cm^2/s; 140 °F (60 °C) [ASTM D445]

2.2 Preparation of the Polymer Modified Binder

SBS modified binder was prepared by adding SBS polymer in the liquid asphalt binder based on the total binder weight. In this study, 4% SBS polymer was added to the base binder. At first, 600 gm of asphalt binder was placed in a quart can and heated in a forced draft oven until it became liquid. After that, it was placed in a melting pot equipped with high shear mixture. The setting temperature and rotation for polymer mixing in high shear mixture was 170 °C and 5000 rpm respectively. The sample was mixed for 30 min at 5000 rpm and 2 min at 8000 rpm. After repeating this process for 2.5 h, data for SBS modified samples were collected.

2.3 Preparation of the RAP binder

RAP binder was extracted and collected using standard test method for quantitative extraction of bitumen from bituminous paving mixtures (ASTM D2172-05) [19] and standard practice for recovery of asphalt from solution using the rotary evaporator (ASTM D5404-03) [20]. Nominal maximum size of the RAP aggregate was ½ inch and 1500gm RAP aggregated was used for obtaining extracted RAP binder.

2.4 Preparation of the Rejuvenated Asphalt Binder

In this study, both bio and aromatic rejuvenator was added with the base binder, polymer modified binder, base binder with 20% RAP binder and polymer modified binder with 20% RAP binder. In these four cases, the dosage of the bio rejuvenator was kept at 5% and 15%, whereas aromatic rejuvenator was maintained at 15% and 35%. To prepare rejuvenated asphalt binder, 50 gm of base binder was poured in an 8 oz tin can and heated on hotplate at 100 °C for 15 min. When it became soft, rejuvenator was added to the base binder according to its weight. Preparation of polymer modified rejuvenated asphalt followed the same procedure as adopted for base binder. For preparing RAP binder induced rejuvenated samples, 10 gm of RAP binder was added to 40 gm of base binder. After that, rejuvenator was added in the same way as before (Fig. 1).

2.5 Fourier Transform Infrared Spectroscopy (FT-IR)

In this study, 4300 handheld FT-IR was used to obtain the spectra. Diamond ATR sensor was used for data collection considering its resistivity in corrosion and scratching and sample thickness [21]. Sample was placed on the sensor area by a spatula. Spectra was obtained using default micorlab PC software provided with the instrument. Before each scan, background spectra was collected to eliminate the effect of surrounding environment. 24 scans were conducted for each sample and their average spectra was collected. Spectra was reported in 650–4000 cm^{-1} region with 4 cm^{-1} resolution. For data collection, sample was collected by a spatula and placed on the surface of the ATR accessory. For each case, a total of 5 samples were scanned. Figure 2 shows the instrumentation and sample collection technique with FT-IR. After collecting all spectra, further analysis was performed on the fingerprint region of the spectra ranges form 650–1800 cm^{-1} [22].

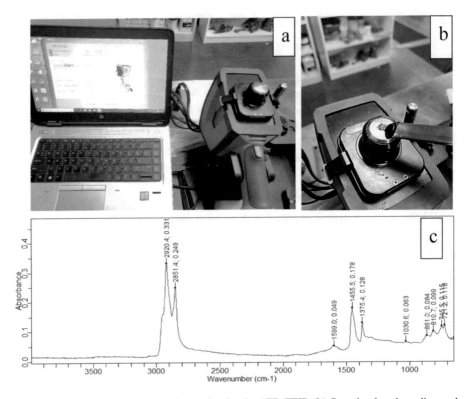

Fig. 1. (a) Instrumental setup for data collection in ATR-FTIR (b) Sample placed on diamond ATR crystal (c) Resultant spectra of the collected sample.

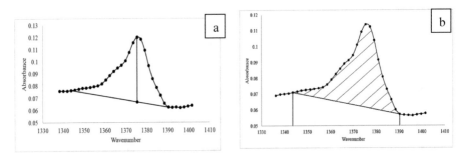

Fig. 2. (a) Absorbance height measurement of FT-IR spectra (b) Absorbance area measurement of FT-IR spectra.

2.6   Analysis Methods

Both qualitative and quantitative analysis were performed to visualize the effect of rejuvenator modification. For qualitative analysis, rejuvenator modified spectra was compared with the standard spectra without rejuvenator modification to identify the presence of unknown substance due to modification. This unknown substance was categorized as characteristic functional group of the sample. After that, a linear relationship was established based on the absorbance intensity and concentration of the rejuvenator. The absorbance intensity was analyzed by three methods (a) absorbance value and (b) absorbance height, and (c) absorbance area. For absorbance value, a calibration curve was produced with the absorbance value at the characteristic wavenumber without any pretreatment of the spectra. It is considered as the simplest method to visualize the linear relationship but highly affected by the shift of the baseline [1]. After that, absorbance height and absorbance area of the characteristic band were calculated. In both cases, a baseline was drawn considering two lowest bands on both side of the definite absorbance band. For absorbance height determination, top point and middle point of the base line was added, and peak height was calculated (Fig. 2a). Absorbance area was calculated using trapezoidal rule. Region of the definite absorbance band was considered as a function of trapezoid and the total area under the band was calculated form the x-axis. After that, area under the baseline was calculated and subtracted from the whole area which give the interested area under the absorbance band (Fig. 2b).

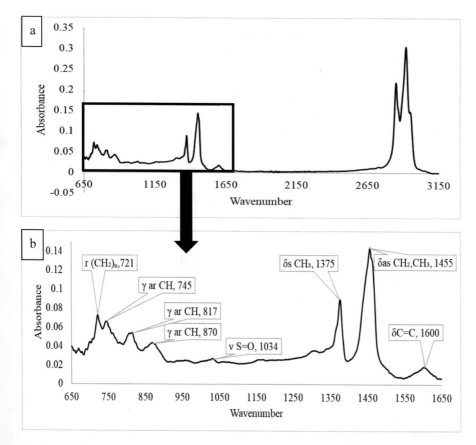

Fig. 3. Absorbance Spectra of Aromatic Rejuvenator in ATR-FTIR (a) spectral region from 650–4000 cm^{-1} (b) Fingerprint spectral region from 650–1800 cm^{-1}

3   Result and Discussion

3.1   Qualitative Analysis

For qualitative analysis neat spectra of the asphalt binder, aromatic rejuvenator and bio rejuvenator were studied to identify their representative functional groups. Figures 3 and 4 showed the representative functional groups present in aromatic and bio rejuvenator. The characteristics functional groups in the fingerprint region of the aromatic rejuvenator are- bending vibration of C=C at 1600 cm^{-1}, asymmetric bending of CH_2 and CH_3 at 1455 cm^{-1}, symmetric bending of CH_3 at 1375 cm^{-1}, stretching vibration of S=O at 1034 cm^{-1}, out of plane deformation CH at 870, 817, 745 cm^{-1} and aromatic ring vibration of CH at 721 cm^{-1}. On the other hand, representative functional groups of the bio rejuvenator in the fingerprint region are stretching vibration of C=O at 1744 cm^{-1}, bending vibration of C=C at 1653 cm^{-1}, stretching vibration of CH_2 and CH_3 at 1453 cm^{-1}, bending vibration of CH at 1377 cm^{-1}, stretching vibration of C-O at 1243, 1162, 1118, 1097 cm^{-1}, and ring vibration of CH_2 at 721 cm^{-1}. The functional

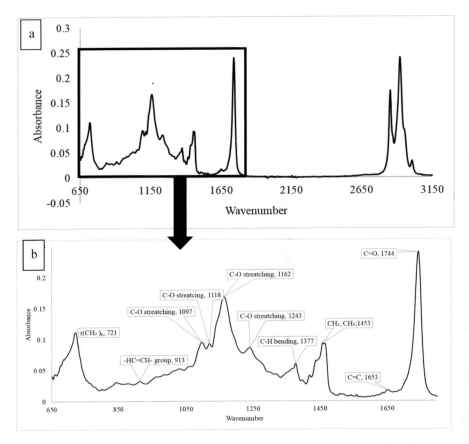

Fig. 4. Absorbance spectra of bio Rejuvenator in ATR-FTIR (a) spectral region from 650–4000 cm^{-1} (b) Fingerprint spectral region from 650–1800 cm^{-1}

groups in fingerprint region of aromatic rejuvenator are same as asphalt binder [18]. When bio rejuvenator added with unmodified asphalt binder, spectra showed two distinct functional groups in the fingerprint region- stretching vibration of C=O at 1744 cm^{-1} and stretching vibration of C-O at 1162 cm^{-1}. These two functional groups were also apparent when bio rejuvenator added in polymer modified binder and RAP induced unmodified and polymer modified binder. From qualitative analysis, it was determined that these two functional groups can be a good representation for identification of bio rejuvenator in asphalt binder. Since, fingerprint region of asphalt binder and aromatic rejuvenator are composed of same functional groups, no distinctive peaks were observed when aromatic rejuvenator mixed with asphalt binder. Figures 5 and 6 showed the effect of bio rejuvenator and aromatic rejuvenator addition in unmodified binder, polymer modified binder and RAP induced unmodified and polymer modified binder.

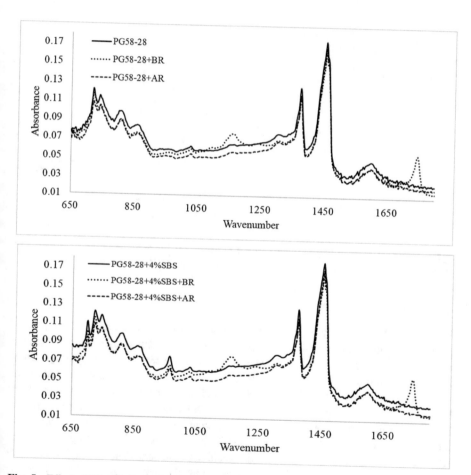

Fig. 5. Effect of BR (Bio Rejuvenator) and AR (Aromatic Rejuvenator) on unmodified and SBS modified binder spectra in 650–1800 spectral region.

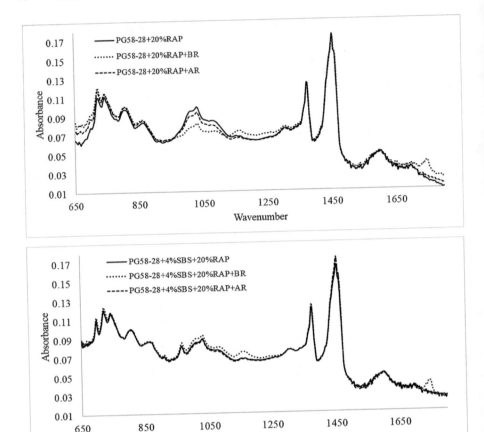

Fig. 6. Effect of BR (Bio Rejuvenator) and AR (Aromatic Rejuvenator) on unmodified and SBS modified binder spectra modified with 20% RAP binder in 650–1800 spectral region.

3.2   Quantitative Analysis

From the qualitative analysis, it was apparent that bio rejuvenator added two functional groups in asphalt binder which can be used for quantification of bio rejuvenator in asphalt binder. The fact that absorbance in ATR-FTIR spectra is linearly proportional to the concentration which directed to conduct a linear regression analysis based on the absorbance intensity, absorbance height and absorbance area at wavenumber 1744 cm^{-1} and 1163 cm^{-1}, responsible for bio rejuvenation. The base lines for height and area measurement were drawn from wavenumber $1764–1720 \text{ cm}^{-1}$ for C=O stretching vibration at 1744 cm^{-1} and $1200–1129 \text{ cm}^{-1}$ for C-O stretching vibration at 1162 cm^{-1}. Tables 2 and 3 showed the results of the regression analysis in these wavenumbers. Results presented that absorbance height method exhibited a good linear relationship with R^2 ranges from 0.97 to 0.99 than the other two methods. It was also found that linear regression based on the stretching vibration of C=O (1744 cm^{-1}) was

Table 2. Properties of linear regression curves due bio rejuvenation process

Functional group	Analysis method	Samples	Regression equation		
			Slope	Intercept	R^2
Stretching vibration of C=O at 1744 cm^{-1}	Absorbance value	UMB+BR	0.0161	0.0082	0.997
		PMB+BR	0.0147	0.0106	0.969
		20%RAP+UMB+BR	0.0056	0.0220	0.979
		20%RAP+PMB+BR	0.0064	0.0216	0.989
	Absorbance height	UMB+BR	0.02	−0.01	0.994
		PMB+BR	0.02	−0.02	0.983
		20%RAP+UMB+BR	0.0055	−0.0055	0.970
		20%RAP+PMB+BR	0.0066	−0.0065	0.990
	Absorbance area	UMB+BR	0.30	−0.24	0.992
		PMB+BR	0.31	−0.30	0.984
		20%RAP+UMB+BR	0.1072	−0.0796	0.968
		20%RAP+PMB+BR	0.1310	−0.1033	0.991

UMB = Unmodified binder; PMB = Polymer modified binder, BR = Bio Rejuvenator, AR = Aromatic Rejuvenator

Table 3. Properties of linear regression curves due bio rejuvenation process

Functional group	Analysis method	Samples	Regression equation		
			Slope	Intercept	R^2
Stretching vibration of C-O at 1162 cm^{-1}	Absorbance value	UMB+BR	0.0054	0.0598	1.000
		PMB+BR	0.0046	0.0612	0.811
		20%RAP+UMB+BR	0.0020	0.0674	0.906
		20%RAP+PMB+BR	0.0027	0.0668	0.990
	Absorbance height	UMB+BR	0.0049	−0.0024	0.9976
		PMB+BR	0.0052	−0.0032	0.9808
		20%RAP+UMB+BR	0.0017	−0.0001	0.968
		20%RAP+PMB+BR	0.0021	−0.0006	0.978
	Absorbance area	UMB+BR	0.11	0.08	0.945
		PMB+BR	0.18	−0.11	0.983
		20%RAP+UMB+BR	0.0627	−0.0083	0.962
		20%RAP+PMB+BR	0.0753	−0.0203	0.978

UMB = Unmodified binder; PMB = Polymer modified binder, BR = Bio Rejuvenator, AR = Aromatic Rejuvenator

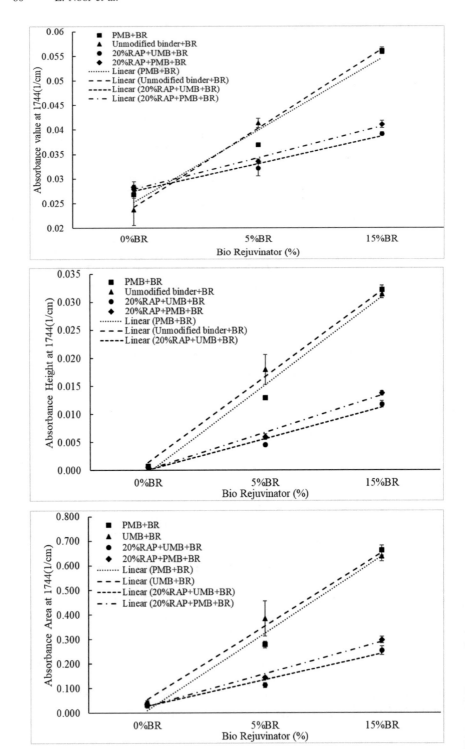

Fig. 7. Linear regression analysis for bio rejuvenated FT-IR spectra at wavenumber 1744 cm^{-1}

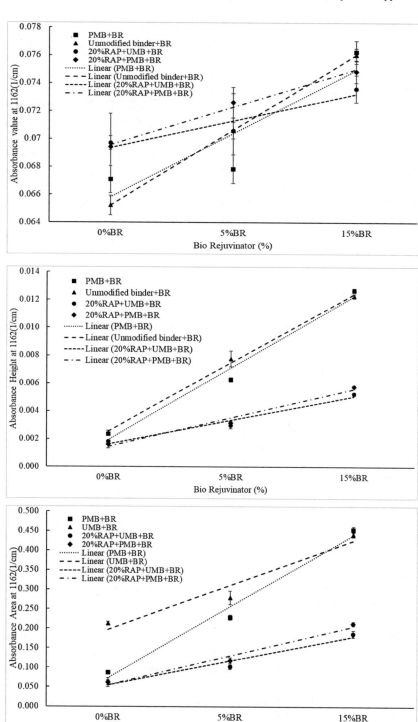

Fig. 8. Linear regression analysis for bio rejuvenated FT-IR spectra at wavenumber 1162 cm^{-1}

more fitted than the stretching vibration of C-O (1163 cm^{-1}). From these two-inference, additional conclusion was carried out based on the regression equation established from absorbance height measurement at wavenumber 1744 cm^{-1}. It suggested that, slope and intercept of the regression line did not vary significantly when bio rejuvenator was added to the unmodified and polymer modified binder, which indicated that quantification of bio rejuvenator did not affect by the presence of SBS polymer. Again, addition of RAP binder caused a significant decrease in C=O functional group as well as slope and intercept of the regression equations. It suggested that quantification of bio rejuvenator can also be supportive to comment on if any RAP binder is added with the unmodified binder or polymer modified binder. A reduction in absorbance intensity of C=O stretching vibration occurs due to the addition of the RAP binder. This was caused by the electron distribution between the stretching vibrations of C=O at 1744 cm^{-1} and the adjacent 1694 cm^{-1}, which was responsible for oxidation in the RAP binder. Figures 7 and 8 showed the regression line at wavenumber 1744 cm^{-1} and 1162 cm^{-1} respectively.

4 Conclusion

This study focused on the performance of the handheld FT-IR as an individual technique in identification and quantification of the bio and aromatic rejuvenator in asphalt binder. Since chemical characterization of the rejuvenator is a demanding phenomenon, FT-IR showed its applicability to differentiate between aromatic and bio rejuvenator by analyzing their functional group incorporated in the asphalt binder. The key findings of this study are listed below.

1. From the qualitative analysis of the FT-IR spectra, it was apparent that bio rejuvenator added two functional groups- stretching vibration of C=O at 1744 cm^{-1} and C-O at 1162 cm^{-1} which can be used to identify bio rejuvenator in asphalt samples.
2. Among three different spectral analysis, absorbance height method showed good linear relationship with a R^2 value of 0.96 in quantifying bio rejuvenator in asphalt binder.
3. Slope and intercept of the regression equations did not vary significantly when bio rejuvenator was added to the unmodified and polymer (SBS) modified binder. This phenomenon can be inferred as the quantification process of the bio rejuvenator having no effect due the presence of polymer.
4. Slope and intercept of the 20% RAP binder induced samples were significantly lower in comparison to when RAP binder was not induced to samples. The reduction in slope and intercept were caused by the reduction in the absorbance height at 1744 cm^{-1} which was affected by the electron distribution form the adjacent oxidizing molecular compound at 1694 cm^{-1}. Although, reduction in absorbance height in the representing bio rejuvenator functional group did not affect the quantification process with a $R^2 > 0.96$.
5. Unlike bio rejuvenator, aromatic rejuvenator did not add any significant functional group in the asphalt binder.

References

1. Hou, X., Lv, S., Chen, Z., Xiao, F.: Applications of Fourier transform infrared spectroscopy technologies on asphalt materials. Measurement **121**, 304–316 (2018)
2. Ren, R., Han, K., Zhao, P., Shi, J., Zhao, L., Gao, D., Yang, Z.: Identification of asphalt fingerprints based on ATR-FTIR spectroscopy and principal component-linear discriminant analysis. Constr. Build. Mat. **198**, 662–668 (2019)
3. Ghauch, A., Deveau, P.A., Jacob, V., Baussand, P.: Use of FTIR spectroscopy coupled with ATR for the determination of atmospheric compounds. Talanta **68**(4), 1294–1302 (2006)
4. DeDene, C.D., You, Z.P.: The performance of aged asphalt materials rejuvenated with waste engine oil. Int. J. Pavement Res. Technol. **7**(2), 145–152 (2014)
5. Jia, X., Huang, B., Bowers, B.F., Zhao, S.: Infrared spectra and rheological properties of asphalt cement containing waste engine oil residues. Constr. Build. Mat. **50**, 683–691 (2014)
6. Azahar, W.N.A.W., Jaya, R.P., Hainin, M.R., Bujang, M., Ngadi, N.: Chemical modification of waste cooking oil to improve the physical and rheological properties of asphalt binder. Constr. Build. mat. **126**, 218–226 (2016)
7. Asli, H., Ahmadinia, E., Zargar, M., Karim, M.R.: Investigation on physical properties of waste cooking oil–Rejuvenated bitumen binder. Constr. Build. Mat. **37**, 398–405 (2012)
8. Sun, Z., Yi, J., Huang, Y., Feng, D., Guo, C.: Properties of asphalt binder modified by bio-oil derived from waste cooking oil. Constr. Build. Mat. **102**, 496–504 (2016)
9. Chen, M., Xiao, F., Putman, B., Leng, B., Wu, S.: High temperature properties of rejuvenating recovered binder with rejuvenator, waste cooking and cotton seed oils. Constr. Build. Mat. **59**, 10–16 (2014)
10. Sun, D., Lu, T., Xiao, F., Zhu, X., Sun, G.: Formulation and aging resistance of modified bio-asphalt containing high percentage of waste cooking oil residues. J. Cleaner Prod. **161**, 1203–1214 (2017)
11. Nguyen, D., Haghshenas Fatmehsari, H., Kommidi, S., Kim, Y.R.: Optimizing Chemical & Rheological Properties of Rejuvenated Bitumen (2016)
12. Zhu, H., Xu, G., Gong, M., Yang, J.: Recycling long-term-aged asphalts using bio-binder/plasticizer-based rejuvenator. Constr. Build. Mat. **147**, 117–129 (2017)
13. Cavalli, M.C., Zaumanis, M., Mazza, E., Partl, M.N., Poulikakos, L.D.: Effect of ageing on the mechanical and chemical properties of binder from RAP treated with bio-based rejuvenators. Compos. Part B: Eng. **141**, 174–181 (2018)
14. Zhang, R., You, Z., Wang, H., Ye, M., Yap, Y.K., Si, C.: The impact of bio-oil as rejuvenator for aged asphalt binder. Constr. Build. Mat. **196**, 134–143 (2019)
15. Cai, X., Zhang, J., Xu, G., Gong, M., Chen, X., Yang, J.: Internal aging indexes to characterize the aging behavior of two bio-rejuvenated asphalts. J. Cleaner Prod. **220**, 1231–1238 (2019)
16. Elkashef, M., Williams, R.C., Cochran, E.W.: Thermal and cold flow properties of bio-derived rejuvenators and their impact on the properties of rejuvenated asphalt binders. Thermochim. Acta **671**, 48–53 (2019)
17. Kaseer, F., Martin, A.E., Arámbula-Mercado, E.: Use of recycling agents in asphalt mixtures with high recycled materials contents in the United States: a literature review. Constr. Build. Mat. **211**, 974–987 (2019)
18. Yut, I., Zofka, A.: Attenuated total reflection (ATR) Fourier transform infrared (FT-IR) spectroscopy of oxidized polymer-modified bitumens. Appl. Spectrosc. **65**(7), 765–770 (2011)

19. ASTM D2172-05, Standard Test Methods for Quantitative Extraction of Bitumen From Bituminous Paving Mixtures, ASTM International, West Conshohocken, PA (2005). www.astm.org
20. ASTM D5404-03, Standard Practice for Recovery of Asphalt from Solution Using the Rotary Evaporator, ASTM International, West Conshohocken, PA (2003). www.astm.org
21. AASTHO T302, Standard Method of Test for Polymer Content of Polymer-Modified Emulsified Asphalt Residue and Asphalt Binders, American Association of State Highway and Transportation Officials (AASHTO) (2015)
22. Marcelo, M.C.A., Mariotti, K.C., Ferrão, M.F., Ortiz, R.S.: Profiling cocaine by ATR–FTIR. Forensic Sci. Int. **246**, 65–71 (2015)

Investigation on the Use of E-Waste and Waste Plastic in Road Construction

Shrikant Dombe[1], Anand B. Tapase[2(✉)], Y. M. Ghugal[3(✉)],
B. A. Konnur[1], and Patil Akshay[1]

[1] Department of Civil Engineering, Government College of Engineering, Karad,
Karad, Maharashtra, India
shrikantdombe@gmail.com, bakonnur@gmail.com,
patil.akshay1595@gmail.com
[2] Department of Civil Engineering, Rayat Shikshan Sanstha's Karmaveer
Bhaurao Patil College of Engineering, Satara, Maharashtra, India
tapaseanand@gmail.com
[3] Applied Mechanics Department, Government College of Engineering, Karad,
Karad, Maharashtra, India
yuwraj.ghugal@gcekarad.ac.in

Abstract. The amount of e-waste and waste plastic caused by technological and industrial development is increasing day by day. One of the low-cost and most effective way to exclude such waste and protect the environment is recycling, wherein its advantageous use in the road construction sector by replacing conventional exhausting materials can be a good eco-friendly alternative. In the initial stage, an attempt to state the potential use of various waste material reported by the number of researchers is done from an extensive literature review. Furthermore, to check cost effectiveness along with sustainable construction method, the experimental investigation on studying the effect of partial replacement of filler and bitumen by e-waste and waste plastic in the bituminous mixes respectively is done by using Marshall Stability testing machine. To check the suitability of e-waste and waste plastic in road construction, the results obtained from the Marshall Stability testing machine for the number of trials were interpreted in respect to its stability, flow value, % V.M.A., % air voids, bulk density, etc. It was observed that partial replacement of conventional material with e-waste is possible which not only the increase in strength but also gives a cost-effective solution towards the e-waste disposal. Advantageous use of plastic is well documented in the available literature and is again confirmed from the present work.

Keywords: E-waste · Partial replacement · Cost-effectiveness

1 Introduction

The rising amount of commercial vehicles, overloading of trucks further than double its capacity, change in daily and cyclic temperature and environmental factors have been responsible for decreasing the life of the pavement. As per the Research Scheme R-55 of MORTH, the use of E-Waste and Plastic waste in bituminous road construction of

S. Badawy and D.-H. Chen (Eds.): GeoMEast 2019, SUCI, pp. 85–99, 2020.
https://doi.org/10.1007/978-3-030-34196-1_6

Central Road Research Institute (CRRI) indicates that the wearing course of e-waste and plastic waste bituminous mixes have longer lives. Indian Road Congress (IRC) has formulated IRC codes for the use of waste plastic and e-waste in road construction. From the literature, it is observed that the properties of pavements with the bituminous mixes can be improved to meet the requirements of pavement with the incorporation of certain modifiers. To achieve this improvement, it is necessary to add polymers to bituminous mixes. Waste plastic is added to enhance the property of the bituminous mixes resulting in improvement of the quality of roads. In India, huge investments are being made to enlarge the existing road infrastructure and also to develop new highways for superior connectivity as well as to serve the economic activity in the country. The time has come to consider these infrastructures as a national asset and to establish a reliable plan to maintain and manage these assets. In India, due to innovation and advancement in new techniques, the electronic equipment's are assembly more attention across the world, due to which modern and most upgraded version is available in the market and the older becomes scrap. The use of these materials as a substitute to conventional material for the construction industry which may not only help in decreasing the manufacturing cost of a particular item but also helps in saving the environment from pollution and other harmful effects which causes problems, reduce landfill cost and also helps in saving our natural resources. The factors like economic growth increased income and the mobility motive has encouraged the people to opt for private vehicles. Also, the inadequate availability, substandard quality, and service of the city's public transportation have further accentuated the trend of using private vehicles in the state. Though there have been considerable improvements in the service of public transport facilities in most of the cities with a growing economy, there still remains a considerable gap between demand and supply of transport facilities.

2 Necessity of Work

Roads in India are largely designed based on soil (sub-grade) strength in terms of California Bearing Ratio and the effect of overloading on the pavement is not considered. The increase in traffic volume and the load intensity cause distress to pavements. In such situations, it becomes imperative to strengthen the pavements by providing overlays, etc. To rehabilitate distress roads, the conventional method of laying overlays is not only time consuming but it also obstructs the flow of traffic for a longer period. The non-availability & increasing cost of conventional granular materials make it necessary to think for better alternatives. So, it is a necessity to find an alternative method to improve the strength properties of layers in flexible pavement meeting the additional requirements put forward due to ever-increasing traffic & other related factors.

3 Aims and Objectives

1. To study the use of electronic waste in the bituminous top layer of flexible pavement along with partial replacement of bitumen with plastic by wet mix process.

2. To find a suitable alternative over conventional materials with cost reduction & improvement in strength & other parameters in flexible pavements.
3. To use waste material in flexible pavement without increasing the cost per unit and without sacrificing durability.

4 Materials and Methods

4.1 Bitumen

Bitumen is defined as "A viscous liquid, or a solid, consisting basically of hydrocarbons and their derivatives, which is soluble in trichloroethylene and is substantially nonvolatile and softens gradually when heated. Bitumen is obtained as the last residue in the fractional distillation of crude petroleum.

4.2 Grades of Bitumen Used for Pavement Purpose

Three grades of bitumen confirming to IS 73:1992 are manufactured in India.

a. Bitumen 80/100: This grade is most extensively used in road construction all over India.
b. Bitumen 60/70: 60/70 grades are harder in assessment to 80/100 and are presently used in the construction of National Highways & State Highways.
c. Bitumen 30/40: Bitumen 30/40 is used in particular applications like airport runway construction as this grade is harder than 60/70 grade.

Sr. no.	Tests conducted	Test results	Specifications	IS codes
1	Specific Gravity	1.02	0.97–1.02	IS-1202
2	Flashpoint, °C	270	220 °C	IS-1209
3	Penetration test	65	50-90	IS-1203
4	Softening point test, °C	49	Min 55 °C	IS-1205
5	Ductility test	51	Min 40	IS-1208

4.3 Aggregate

Aggregates are natural materials and are the most significant constituents of road construction. Aggregates are also used as foundation material under road construction. The most universal natural aggregates of mineral origin are sand, gravel, and crushed rock. Unbound aggregates are used for base or sub-base course. Aggregate contributes up to 90–95% of the mixture weight and contributes to most of the load bearing & strength characteristics of the mixture. There are three kinds of rocks namely-igneous, sedimentary, and metamorphic. It may found that many of the properties of aggregates namely-specific gravity, hardness, strength, physical stability depend mostly on the quality of parent rock.

Physical Requirements for Coarse Aggregates for Bituminous Concrete Table 500-16.

Property	Test	Specifications	Method of test
Cleanliness (Dust)	Grain size analysis	Max 5% passing 0.075 mm sieve	IS:2386 Part 1
Particle shape	Combined Flakiness and Elongation Indices	Max 35%	IS:2386 Part 1
Strength	Los Angeles abrasion value/Aggregate Impact value	Max 30% Max 24%	IS:2386 Part 4
Durability	Soundness either: Sodium Sulphate or Magnesium Sulphate	Max 12% Max 18%	IS:2386 Part 5
Polishing	Polished Stone Value	Min 55	BS:812-114
Water absorption	Water absorption	Max 2%	IS:2386 Part 3
Stripping	Coating and stripping of Bitumen-Aggregate Mixtures	Min retained coating 95%	IS:6241
Water sensitivity	Retained Tensile Strength	Min 80%	AASHTO 283

4.4 E-Waste

There is no clear definition for electronic waste (e-waste) at this time, but if you can plug it into an electrical outlet or it contains circuit boards or chips, it is most likely e-waste. This term applies to customer and business electronic equipment that is near or at the end of its useful life. Electric and electronic waste (e-waste) is currently the largest growing waste stream. It is harmful, composite and luxurious to treat in an environmentally sound manner and there is a general lack of legislation or enforcement surrounding it. Today, most e-waste is being surplus in the general waste stream. These products can contain heavy metals like cadmium, lead, copper, and chromium that can contaminate the environment.

Examples of electronic waste include, but not limited to:

a. TVs, computer monitors, printers, scanners, keyboards, mice, cables, circuit boards, lamps, clocks, flashlight, calculators, phones, answering machines, digital/video cameras, radios, VCRs, DVD players, MP3, and CD players
b. Kitchen equipment (toasters, coffee makers, microwave ovens)
c. Laboratory equipment (hot plates, microscopes, calorimeters)
d. Broken computer monitors, television tubes (CRTs)

4.5 Plastic

Safe disposal of waste plastic is a serious environmental problem. Being a non-biodegradable material it does not decay over time and even if dumped in landfills, finds its way back in the environment through air and water erosion can obstruct the drains and drainage channels. Bottles, containers and packing strips, etc. is increasing day by day. As a result of the amount of waste plastic also increases. This leads to various environmental problems. Many of the wastes produced today will remain in the environment for many years leading to various environmental concerns. Therefore it is necessary to utilize the wastes effectively with technical development in each field. Many by-products are being produced using plastic wastes. Our present work is helping to take care of these aspects. Plastic waste, consisting of carrying bags, cups and other utilized plastic can be used as a coating over aggregate and this coated stone can be used for road construction. The mix polymer coated aggregate and tire modified bitumen have shown higher strength. The use of this mix for road construction helps to use plastics waste. Once the plastic waste is separated from municipal solid waste, the organic matter can be converted into manure and used. Our paper will discuss in detail the process and its successful applications.

Plastic is a very adaptable material. Due to the industrial revolution, and its large scale production plastic seemed to be a cheaper and effective raw material. Today, every vital sector of the economy starting from agriculture to packing, automobile, electronics, electrical, building construction, and communication sector has been virtually revolutionized by the application of plastics. Plastic is a non-biodegradable material and researchers are found that the material can remain on earth for 4500 years without degradation. We cannot ban the use of plastic but we can reuse plastic waste.

Plastics and their origin

Waste plastic	Origin
Low-density polyethylene (LDPE)	Carry bags, sacks, milk pouches, bin lining, cosmetic and detergent bottles, etc.
High-density polyethylene (HDPE)	Carry bags, bottle caps, household articles, etc.
Polystyrene Terephthalate (PET)	Bottle caps and closures, film wrappers of detergent, biscuit, wafer packets, microwave trays for a ready-made meal, etc.
Polypropylene (PP)	Yogurt pots, clear egg packs, bottle caps. Foamed Polystyrene: food trays, egg boxes, disposable cups, protective packaging, etc.
Polystyrene (PS)	Food trays, egg boxes, disposable cups, protective packaging, etc.

Tapase Anand et al., "Utilization of E-Waste and Polymer Modified Bitumen in Flexible Pavement"

The escalation in various types of productions together with population growth has resulted in a massive increase in the production of various types of the waste material world over creating a problem of its disposal in an eco-friendly way. To deal with the problem here an attempt is made to study the use of e-waste as an alternative to conventional material like aggregate in a DBM layer of flexible pavement along with partial replacement of bitumen with plastic by the wet mix process. A number of laboratory tests were conducted using the Marshall Stability testing machine to check the suitability of e-waste and plastic as an alternative to conventional materials like aggregates and bitumen respectively. The results obtained in laboratory investigation indicate not only the increase in strength but also a considerable reduction in cost is seen. From the experimental work, it is clear that the properties of laboratories designed bituminous mix for DBM are much more superior to those of the control mixes entirely composed of mineral aggregates and can be effectively used in practical applications.

Tapase Anand, et al., "Consumption of Electronic Waste in Quality Enhancement of Road"

The work consists of an experimental approach towards waste management and finding an alternative to conventional materials in flexible pavements. Most of the electronic waste is recyclable or repairable, but a number of worthless electronic pieces causes higher transportation cost for their processing which may be higher than its scrap value. So, such electronic waste is disposed of very casually, which may cause serious health and pollution problems. Also, the disposal of electronic waste is difficult because of non-degradable plastic contents and metals like lithium, copper, and aluminum, which may lead to adverse effects on the environment. To deal with the problem, here an attempt is made to study the use of electronic waste as an alternative to conventional material like aggregate in a DBM layer of flexible pavement. A number of laboratory tests were carried out by replacing aggregates partially by shredded electronic waste. The outcomes from the laboratory investigation prove the suitability of electronic waste in road construction with substantial cost saving. So, disposal of hazardous electronic waste in the pavement can prove to be one of the alternatives to make the earth greener and pavements more durable.

S. Rajasekaran et al., "Reuse of Waste Plastic Coated Aggregates- Bitumen Mix Composite for Road Application-Green Method"

Normally Waste plastic is made up of PP, PE and P. Softening point of plastic up to 130 °C and they do not produce any toxic gases during the softening condition. The nature of softened plastic is to form a coating like material over the material. Spray of plastic over the hot aggregate at 160 °C to form the film-like structure on the aggregate. It formed PCA, which is a better raw material for the construction of the bitumen road. Moisture absorption test, Soundness test, Aggregate impact test, Los Angeles abrasion test conducted in this paper. Mix the plastic coated aggregates in 80/100 pen bitumen at 160 °C after the completion of test Mix. The researcher took Marshall Stability tests for bitumen. The mixture of polyethylene coated aggregate and bitumen mix shows the improvement in binding property of aggregate and less wetting property. The control mix with plastic sample shows greater Marshall Stability value is up to 18–20 KN also increases the load-bearing capacity up to 100%.

Tapase Anand et al., "Performance Evaluation of Polymer Modified Bitumen in Flexible Pavement"

The growth in various types of industries together with population growth has resulted in an enormous increase in the production of various types of waste materials world over creating a problem of its disposal in an eco-friendly way. To deal with the problem, a study on the use of plastic waste as a partial replacement to bitumen in the flexible pavement is considered in the present work. The work consists of an experimental approach towards waste management and finding an alternative to conventional materials in flexible pavements. To simulate with the field conditions the Marshall Stability method is considered to carry out experimental work. The objective of work is to investigate the effect of plastic waste in flexible pavement and to suggest the optimum percentage of bitumen that can be replaced by plastic waste for the improvement of roads. The number of laboratory tests has been carried out by replacing bitumen with plastic waste. The results obtained in laboratory investigation indicate a major gain in strength with substantial savings in cost.

M.S. Ranadive et al., "Performance Evaluation of E-Waste In Flexible Pavement – An Experimental Approach"

The objective of this study is to investigate the effect of e-waste and fly ash, as a filler replacement, on the strength parameters of the bituminous concrete road. It is observed that there is a definite increase in Marshall Stability and flows value with an increase in the percentage of e-waste by replacing up to 10 percent aggregates. The use of fly ash as a filler cannot increase the strength but helps to attain it nearly equal to that of the control mix. Here is an attempt is made to use waste products like fly-ash and e-waste in flexible road construction which affect the environment and are difficult to process. The study concludes that e-waste and fly ash could be used as filler material in bituminous mixes.

Silvia Angelone et al., "Green pavements: reuse of plastic waste in asphalt mixtures"

The important objective of this research is to find an eco-friendly approach for effect of recycling various percentages of urban and rural plastic waste by adding them in asphalt mixture by the dry process, for a comparative laboratory study. Marshall Stability, Indirect tensile strength Marshall Quotient, fracture energy, permanent deformation, creep compliance and resilient modulus, this all factors is conducted for the laboratory study. Polyethylene from silo bags and Polypropylene from waste bags bottles are used as the waste plastic material for the modified bitumen mixture. The proportion of plastic dosage in bitumen is 2%, 4%, 6% weight of the mixture. After conducting all the above tests, researchers concluded that the addition of plastic in flexible pavement improves the properties of flexible pavement.

Zahra Niloofar Kalantar et al.," A review of using waste and virgin polymer in the pavement"

Researchers describe the adding of virgin polymer in bitumen for improving properties of bitumen. More additions of the virgin polymer have the same result as compare to waste polymer according to historical study. This paper review of the virgin also wastes polymer in the pavement with help of study on the history of use of polymer in asphalt, benefits of using polymer in asphalt, use of polymer waste into the virgin polymer. A benefit of using polymer in asphalt section describes the major improvement in Optimum stability of bitumen, Improvement in resistance of bitumen temperature, Improvement in engineering properties. The results of the Virgin polymer as compare to waste polymer are the same. So the economic and environmental point of view the use of waste polymer is the best modifier as compare to virgin polymer I.e. waste polymer solves the waste-disposal problem also improves in performance of the pavement. Researchers give more information related to polymer characteristics and bitumen characteristics. After all review, researchers concluded that the use of virgin polymer in asphalt pavement is better for improving in the certain characteristics of asphalt pavement. From an economic view and environmental view, the use of waste polymer in asphalt pavement is better.

Amit Gawande et al., "An overview on waste plastic utilization in asphalting of roads"

This paper describes the use of plastic in bitumen to improve desired mechanical characteristics for the flexible road. This paper review important techniques for using waste plastic in construction flexible pavements and bituminous road. Researchers collect more data on plastic consumption and the generation of plastic waste. Also, this paper describes the properties and characteristics of bitumen and plastic. This paper gives detail information about the wet and dry process. The final conclusion of that paper is modified bitumen is better for the upper layer of flexible pavement, and modified bitumen shows more resistant to water, stability, load carrying capacity and better binding property.

Sui Yuanyuan et al., "Application and Performance of Polyethylene Modifying Additive in Asphalt Mixture"

This paper deals with the mixing or adding method of polyethylene modifying additive in different from asphalt modifier. The method of research is mixing the Polyethylene as modifying additive in the mineral aggregate for some minutes and then adds the asphalt mixing in polyethylene and aggregate. This method is different from the regular asphalt modifier. This paper mainly described the improvement in asphalt mixture when it is mixed with polyethylene modifying additive on higher temperature stability, resistance at low temperature cracking and water resistance is obviously, analyze the mechanism that additive affect asphalt mixture performance and evaluate polyethylene modifying additive on the basis of technical, economic and environmental aspects. Also, researchers explained about High-temperature Stability and Low-temperature Performance, Water Stability is weak in bitumen. The conclusion of this research paper is improving the quality of bitumen. (i.e. maximum temperature stability and low-temperature performance, water stability) by mixing polyethylene modifying additive.

5   Wet Mix Process

Waste plastic is mixed with bitumen in 4.5% to 6% as partial replacement to bitumen. Aggregates were partially replaced by the electronic waste in a shredded form with 7.5%, 10%, 12.5%, and 15% by volume of the mold. Plastic increases the melting point of bitumen and makes the road retain its flexibility during winters resulting in its long life. Use of shredded plastic waste in road act's as a strong binding agent for making it durable. The plastic waste is melted and mixed with bitumen in a particular ratio. Normally, blending takes place when the temperature reaches 45.5 °C, but when plastic is mixed, it remains stable even at 55 °C. The vigorous tests at the laboratory level proved that the mixes prepared using the treated bitumen binder fulfilled all specified Marshall Mix Design criteria for the surface course of road pavement. There was a substantial increase in the Marshall Stability value of mix, of the order of "two to three times" higher value in comparison to ordinary, untreated bitumen.

6 Bitumen Tests

Bitumen is an important parameter for flexible pavement; the binding property of bitumen is played an important role in flexible pavement. Plastic waste is mixed in bitumen by the wet mix process. Modified bitumen is the partial replacement of bitumen by waste plastic. Replacement quantity of waste plastic is 6.5% of bitumen weight as per IRC: SP: 53-2010. In the wet mix process, Temperature parameter is important for all process.

6.1 Wet Process

Plastic has a high melting point also increases the melting point of the modified bitumen and makes the road retain its flexibility during winters resulting in its long life. Use of shredded plastic waste which is passing through 4.75 mm sieve and mix with bitumen at 160 °C to 165 °C. Addition of 6.5% waste plastic in bitumen as per IRC: SP: 98-2013.

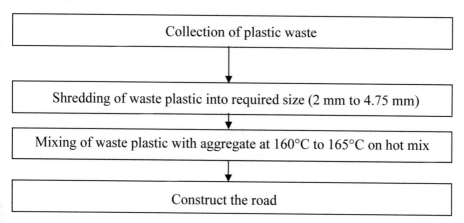

6.2 Advantages of Wet Process

This Process is applied for recycling of any type, size, shape of waste material (Plastics, Rubber, etc.).

6.3   Disadvantages of Wet Process

Required more time- high energy for blending, a special type of equipment is required. Extra time required for cooling.

7   Mix Design

The gradations adopted for the mix are taken as specified by MORTH 2001. Here mix design is made for Bituminous top layer of flexible pavement, in which aggregates size of 20 mm, 12.5 mm, 6 mm, and grit are used. After blending operation, for grade-I, it is observed that the percentage of aggregates mentioned above is 25%, 25%, 20%, and 30% respectively.

Following steps are carried out while preparing mix design:

1. Sieve analysis
2. Mix gradation according to MORTH 2001
3. Mix design blending
4. Recommendations for mix design from the overall process

Table 500-17: Composition of bituminous concrete pavement layer

Grading	I
Nominal aggregate size	19 mm
Layer thickness	50 mm
IS Sieve (mm)	Cumulative % by weight of total aggregate passing
45	–
37.5	–
26.5	100
19	90–100
13.2	59–79
9.5	52–72
4.75	35–55
2.36	28–44
1.18	20–34
0.6	15–27
0.3	10–20
0.15	5–13
0.075	2–8
Bitumen content % by mass of total mix	Min 5.2
Bitumen grade (Penetration)	65 to 90

8 Marshall Stability Test

This is the most regular method that eventually proved to be the Marshall Stability method developed by Bruce G. Marshall of the Mississippi state highway department. The method is modified and standardized by many organizations, such as ASTM, AASHTO, US Corps of engineers and British Standers Institution. The Ministry's specifications and Indian Road Congress also refer to this method.

The experimental set-up for Marshall Stability test consists of the following:

a. Loading machine (Marshall testing machine)
b. Mould
c. Automatic or manual compaction hammer (4.5 kg)
d. Flowmeter
e. Breaking head
f. Digital water bath
g. Hot plates
h. Ovens
i. Miscellaneous equipment

9 Sample Preparation

Following factors has to be considered while sampling preparation:

1. Steps for sample preparation for control mix with different bitumen content
2. Bitumen replaced by plastic waste
3. Aggregate replaced by e-waste

10 Steps for Sample Preparation for Control Mix with Different Bitumen Content

The specimens for control mixes were prepared. Numbers of samples were prepared by varying bitumen percent from 4.5% to 6% with an increment of 0.5%.
The further process is as given below.

a. Aggregates are oven dried at 105 °C to 110 °C and sieved into specified sizes.
b. About 1200 g aggregates were taken as per mix design proportion is given in Table 1
   And mix thoroughly.
c. Then bitumen is heated up to 160 °C and mixed by weight of aggregate.
d. First, the entire mold was filled and 75 blows were given on either side of the specimen with the manual compactor.

11 Bitumen Replaced by Plastic Waste

For the preparation of sample the steps a. and b. mentioned above are the same. The further process is as discussed below:

c. Before mixing aggregate, the shredded plastic waste was mixed with hot bitumen at 160 °C. The percentage variation in plastic was 4.5%, 5%, 5.5%, 6% of bitumen.
d. Required percent of bitumen was added to the sample and mixed at 160 °C. Molds were filled with bituminous mix and 75 blows were given on either side

This constituted the specimen.

12 Aggregates Replaced by E-Waste

a. Aggregates are oven dried at 105 °C to 110 °C and sieved into specified sizes.
b. Then the aggregates were partially replaced by e-waste by its volume. The e-waste percentage was varied from 7.5% to 15% with an increment of 2.5%. This quantity was mixed with hot aggregate at 160 °C. Then the hot bitumen was added to it.
c. Molds were filled with a modified mix and 75 blows were given on either side which constituted the specimen.

13 Result

(1) In the wet mix process for modified bitumen testing 6.5% accurate percentage value for replacement of bitumen by waste plastic. Penetration value is decreased by 6.68% after mixing of 6.5% waste plastic in bitumen but softening point of bitumen is increased by 8.60% and other properties of bitumen are same.
(2) In Dry mix process for aggregate testing, 7% waste plastic quantity is accurate as compare to another percentage variation. 7% optimum bitumen content is replaced by plastic waste; it forms the plastic coated aggregate.
(3) Specific gravity of plastic coated aggregate is increased by 2.88% after coating of 7% waste plastic on aggregate also crushing value, impact value, loss abrasion value are decreased by 3% to 4%.
(4) This all result of aggregate test indicates that replacement of bitumen by plastic waste is increasing the properties of aggregate.
(5) Plastic coated aggregate increase the stones improving surface property of aggregates.
(6) Plastic coating form thin layer around the aggregate to fulfill parameter and binding property of aggregate. Plastic coated aggregate is used for good performance of flexible pavement.
(7) Marshall Stability value of bituminous material increased in dry mix process and wet mix process.

(8) In wet mix process 6% plastic content is optimum shows the better result as compare to other content.

(9) In Dry mix process plastic content is 7%, shows high Marshall Stability value. Due to addition of waste plastic, Marshall Value increase as compare to control mix. High Marshall Values shows high strength, high durability, and high load carrying capacity. This all parameter shows increase the properties of flexible pavement.

(10) As compare to wet mix process, dry mix process is suitable method for mixing of waste plastic in bitumen for construction of flexible pavement because wet mix process required more time and energy for blending; new equipment's are required for wet mix process, Dry mix process is simple and feasible.

The project work consists of an experimental approach towards waste management and finding alternative to conventional materials in flexible pavements. The objective of work is to investigate the effect of plastic waste and electronic waste in DBM layer flexible pavement, and to suggest the optimum percentage of bitumen that can be replaced by plastic waste for the improvement of roads. Also, aggregates are also replaced by e-waste. Marshall Stability test setup is used for testing 100 mm diameter specimen, to simulate actual field condition. Different performance parameters of specimen are studied for varying percent of e-waste and bitumen content.

Following are some conclusions drawn during testing of specimen and interpretation of results:

a. The use of bitumen with the addition of processed waste plastic of about 5% by weight of bitumen helps in substantially improving the Marshall stability, strength, fatigue life and other desirable properties of bituminous concrete mix, with marginal saving in bitumen usage.

b. Using the wet process with varying percentage of 7.5, 10, 12.5, and 15 of aggregate can be replaced by e-waste and 5% plastic to form modified bitumen in DBM layer having 5.5 percent optimum bitumen content.

c. The process is environment friendly.

d. From the experimental work it is clear that the properties of laboratorial designed bituminous mix for DBM are much more superior to those of the control mixes entirely composed of mineral aggregates and can be effectively used in practical applications.

Thus we can conclude in general that it is a simple process, helps to save cost of bitumen, improves performance of roads, solves problem of plastic waste disposal and it is one time investment for shredding machine which can be reused.

14 Suggestions

a. Instead of going through the banning of plastics, it is important that needed education is to be given.

b. Domestic plastic waste need to be separated at the source & collected efficiently.

c. Awareness camps should be conducted.

d. Electronic waste should not be dumped in landfills.
e. Electronic waste should be efficiently collected.

15 Future Scope

a. e-waste disposal in eco-friendly way is a problem and there is a scope to study e-waste as an alternative to aggregates in road construction.
b. Waste plastic carry bags were used in this project which shows adhesion properties in their molten state hence, there is a scope to use different forms of polymers for road construction.
c. Here, as pavement is an alternative for disposal of plastic waste and e-waste, various other engineering structures can also be analyzed for finding an alternative for waste disposal.

References

1. Tapase Anand, Kadam Digvijay, Mujawar Sahil: Utilization of E-Waste and Polymer Modified Bitumen In Flexible Pavement
2. Rajashree, T., Digvijay, K., Anand, T.: Consumption of Electronic Waste in Quality Enhancement of Road
3. Rajasekaran, S., Dr, R., Vasudevan, S.P.: Reuse of waste plastic coated aggregates- Bitumen mix composite for road application-green method. Am. J. Eng. Res. **108**, 12–18 (2013)
4. Tapase A.B., Kadam D.B.: Performance Evaluation of Polymer Modified Bitumen in Flexible Pavement
5. Ranadive, M.S., Shinde, M.K.: Performance Evaluation of E-Waste In Flexible Pavement – An Experimental Approach
6. Angelone, S., Casaux, M.C., Borghi, M., Martinez, F.O.: Green pavements: reuse of plastic waste in asphalt mixtures. Mater. Struct. **12**(3), 1655–1665 (2016)
7. Niloofar, Z., Kalantar, M.R., Karim, A.M.: A review of using waste and virgin polymer in the pavement. Constr. Build. Mater. **84**, 315–319 (2015)
8. Gawande, A., Zamarea, G., Renge, V.C., Tayde, S., Bharsakale, G.: An overview on waste plastic utilization in asphalting of roads. JERS 17–22 (2012). E-ISSN0976-7916
9. Sui, Y., Chen, Z.: Application and performance of polyethylene modifying additive in Asphalt mixture. In: ICTE 2011 © ASCE, pp. 3–9 (2011)
10. Gupta, S., Veeraragavan, A.: Fatigue behavior of polymer modified bituminous concrete mixtures. J. Indian Roads Congr. 55–64 (2009)
11. Murphy, M., O'Mahony, M., Lycett, C., Jamieson, I.: Recycled Polymers for use as bitumen modifiers. J. Mater. Civ. Eng. **13**(4), 306–314 (2001)
12. Al-Hadidy, A.I., Yi-Qui, T.: Effect of polythene on the life of flexible pavements. Constr. Build. Mater. **23**(3), 1456–1464 (2009)
13. Vasudevan, R., Nigam, S.K., Velkennedy, R., Ramalinga Chandra Sekar, A., Sundarakannan, B.: Utilization of waste polymers for flexible pavement and easy disposal of waste polymers. In: Proceedings of the International Conference on Sustainable Solid Waste Management, Chennai, India, 6–9 September 2007 (2007)

14. Archana, M.R., Satish, S.H., Ashwin, M., Hunashikatti, H.: Effect of waste plastics utilization on indirect tensile strength properties of semi-dense bituminous concrete mixes. Indian Highw. **42**(2), 69–78 (2014)
15. IRC: SP: 98-2013 Guidelines for the use of waste plastic in hot bituminous mixes (dry process) in wearing courses
16. IRC SP-53:2010: Guidelines on the use of modified bitumen in road construction

Behavior of Unpaved Roads Under Different Conditions

Adelino Ferreira[1(✉)], M. I. M. Pinto[2], Arminda Almeida[1],
and Natacha Rodrigues[3]

[1] CITTA, Department of Civil Engineering, University of Coimbra,
Coimbra, Portugal
{adelino,arminda}@dec.uc.pt
[2] CEMMPRE, Department of Civil Engineering, University of Coimbra,
Coimbra, Portugal
isabelmp@dec.uc.pt
[3] Department of Civil Engineering, University of Coimbra, Coimbra, Portugal
rodrigues.ncm@gmail.com

Abstract. This paper presents the results of a laboratory study performed on unpaved roads. Both unreinforced wearing layers and layers reinforced with geosynthetics were considered for their construction and study, as geosynthetics are reported to improve the behavior of the granular layers. Reinforcement was introduced by means of a polyester geogrid. Four different types of soils were considered in the study, each one with different percentages of sand, silt and clay. The soil layers were placed, compacted and tested with the optimum water content. Traditional soil characterization was performed, which included particle size analysis, density of solid particles, consistency limits, compaction and California Bearing Ratio. Wheel Tracking tests were performed to study the unpaved road behavior in the different conditions. In the majority of the tests, a compressive material was used beneath the soil layer, to simulate the existence of a weak and deformable subgrade. The results of the study conducted to the following main conclusions: soil D, the one with the highest percentage of fines, shows the best performance under the Wheel Tracking test, for all conditions under study; soil A, the coarsest soil, performed worst; soils B and C improve their performance significantly when the geogrid reinforcement was included.

1 Introduction

Unpaved road behavior is generally associated with high deformability and stability problems. Most of these roads are not built to withstand the daily traffic of trucks and large agricultural machines or vehicles. These roads are often built on weak subgrades, that is, with very low carrying capacity, which aggravates the problem. Great care should therefore be taken in their design and during construction to minimize this type of problem to provide greater durability and therefore fewer maintenance interventions.

These roads allow the access of the population between a rural and an urban area, and that makes this infrastructure a very important issue in the development of the region and the country. According to data from CIA@2015, unpaved roads account for

© Springer Nature Switzerland AG 2020
S. Badawy and D.-H. Chen (Eds.): GeoMEast 2019, SUCI, pp. 100–106, 2020.
https://doi.org/10.1007/978-3-030-34196-1_7

an important percentage of the roads, not only in developing countries, but also in so-called developed countries. South America (84%), Africa (77%) and Oceania (61%) are the three continents/subcontinents with the highest percentage of unpaved roads. Within each continent/subcontinent, there are sometimes large differences from country to country, with Africa and South America being the most homogeneous. In Europe the difference is large, with the Eastern European countries having the highest percentage of unpaved roads (Estonia: 82%, Latvia: 80%, Hungary: 92%, Macedonia: 32%, Poland: 32%, Belgium: 22%, Finland: 36%, Norway: 19%, Portugal: 14%). Improvement is therefore a very important issue for unpaved roads on weak soils. There are several soil improvement and reinforcement techniques available, and reinforcement with geosynthetic layers is one of them (Tang et al. 2015, Wu et al. 2015, Calvarano et al. 2017, Reilly and Nell 2018). Geotextiles and geogrids are the most commonly used materials for soil reinforcement. They increase soft soil bearing capacity, reduce fill lateral deformation, provide a more favorable stress distribution and reduce vertical deformation due to the membrane effect (IGS@2019). As the depth of the ruts increases, the deformed shape of the geosynthetic provides further reinforcement due to the membrane effect.

The study described here investigates de potential of the Wheel Tracking Machine to simulate the traffic on unpaved roads and investigates the effect of the subgrade strength and the inclusion of the reinforcement on the deformation (rut depth) for different types of soils.

2 Laboratory Program

2.1 Materials

Four different soils were studied (A, B, C and D), and these were obtained by mixing three soils already available in the laboratory in different proportions (Table 1), in order to test a total of four different types of soils and to achieve a more progressive granulometric content variation for those soils. Traditional soil characterization was performed, which included particle size analysis, density of solid particles, consistency limits, compaction (heavy Proctor) and California Bearing Ratio (CBR). The results are given in Table 2.

Table 1. Mixture composition

Tested soils	Soils available in the laboratory		
	Lean clay	Silty sand	Sand
A	–	50%	50%
B	25%	50%	25%
C	25%	75%	–
D	50%	50%	–

Table 2. Soil properties

Properties	Soil A	Soil B	Soil C	Soil D
Density of solid particles	2.75	2.76	2.79	2.78
Liquid limit (%)	–	–	–	23.2
Plasticity limit (%)	–	–	–	21.6
Maximum dry unit weight (kN/m^3)	19.9	20.6	20.6	19.6
Optimum water content (%)	8.1	7.6	8.8	10.2
California Bearing Ratio (%)	6.3	4.4	3.8	3.5
%Sand	91.5	79.9	78.6	64.6
%Silt	7.1	18.6	19.6	31.8
%Clay	1.4	1.5	1.8	3.6
Classification (USCS: ASTM D2487-09)	Sandy silt	Sandy silt	Sandy silt	Silty sand
Classification (AASHTO: ASTM D3282-09)	A1b	A2-4	A2-4	A4

The properties of the reinforcement are given in Table 3. A sponge, placed beneath the soil specimens, was used throughout the study to behave as a compressive material to simulate weak and deformable subgrades for the unpaved road. This sponge is 11.05 mm thick and it is of the same size of the base of the test box, measuring 375 mm × 305 mm. This box is 98 mm high and it is made of steel. These are the maximum dimensions possible to fit the test box in the testing machine, which is a Wheel Tracking Machine.

Table 3. Reinforcement properties of Polyester Secugrid 200/40 R6

Aperture (mm × mm)	Tensile strength (kN/m)	Tensile strength at 5% elongation, machine direction (kN/m)
71 × 25	200/40	140

2.2   Specimen Preparation, Test Apparatus and Test Procedure

Three different specimen arrangements were studied: soil alone; soil with the sponge resting at the bottom of the test box, i.e., beneath the soil; and soil, the sponge beneath the soil, and the reinforcement between the sponge and the soil. The reinforcement was placed in such a way that the longest apertures and lower tensile strength of the geogrid were oriented in the longitudinal direction i.e., the same direction of the wheel movement. The soil was prepared with its optimum water content and it was placed in the test box in 3 layers. Each layer was placed, and then densified by a 10 kN load applied by a compression machine, a Servosis Model ME-402, in order to achieve the same dry unit weight as the maximum value obtained in the compaction tests. The final thickness of each layer was about 3 cm.

The main tests were performed on the Wessex S967 Wheel Tracking Machine, which simulates traffic cycling load and allows the measurement of the deformation of the soil surface (rut depth) caused by the passage of a wheel. All of these tests were carried out with the soil still with its optimum water content. The Wheel Tracking Machine is mainly used to test bituminous pavements and it is set to limit the wheel pass cycles to 10,000 and the deformation to 20 mm (EN 12697-22, 2007). In this study the limit was set to 3,000 cycles, because it was found to be enough to reach a constant deformation.

3   Results and Analysis

3.1   Influence of the Subgrade Characteristics

Tests were performed with and without the compressive material beneath the soil layer, simulating a weak and a strong subgrade soil under the unpaved road, respectively. These results are presented in Fig. 1 and they confirm what would be expected. Indeed, the strength of the subgrade soil has a great influence on the deformation behavior of the unpaved road. For all types of soils (A, B, C and D), rut deformation increases significantly with the inclusion of the sponge, with soil A being the most affected soil. Soil A is the soil with less content of fines and therefore with lower capacity to aggregate the particles, which might justify the observed behavior.

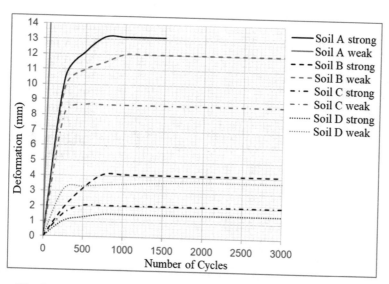

Fig. 1. Rut deformation - unreinforced soil on strong and weak subgrade

For soil A, when comparing results for just 72 cycles, the deformation increased from 3 mm to 20 mm. At 20 mm the test was stopped because deformation reached the maximum value possible for the equipment. This corresponds to an increase in deformation as high as 566.7%. Soil D is the soil with lowest level of deformation for both situations, weak and strong subgrade. Furthermore, it is the one that was least affected by the characteristics of the foundation, in comparison to soils B and C, with just 146.7% aggravation of deformation (Table 4).

Table 4. Influence of the subgrade strength on rut deformation

Soil	Strong subgrade		Weak subgrade		Deformation variation (%)
	Deformation (mm)	Number of cycles	Deformation (mm)	Number of cycles	
A	13.14	1500	20.0*	72	n.a.
B	4.11	3000	12.04	3000	+192.9
C	2.06	3000	8.69	3000	+321.8
D	1.5	3000	3.7	3000	+146.7

* Test stopped at the end of 72 cycles as the safety limit value for deformation was reached.

3.2   Influence of the Inclusion of Reinforcement

To study the influence of the reinforcement, tests were performed with the soil, the sponge and the reinforcement (geogrid) placed with the higher tensile strength and narrower apertures in the transversal direction of the road. From bottom to the top, these materials were placed in the following order: sponge, reinforcement, soil. The results of these tests are plotted in Fig. 2 and some important values are given in Table 5.

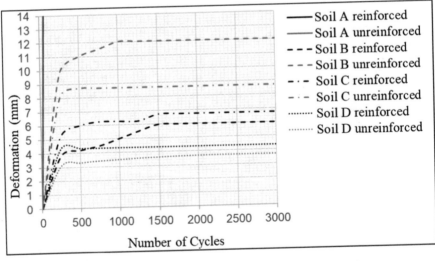

Fig. 2. Rut deformation - reinforced soil on weak subgrade

Reinforcement does not seem to be effective in reducing deformation, neither for the coarsest soil (soil A) nor for the finest soil (soil D). Nevertheless, soil D is the soil that presents the least deformation for all tested conditions. Reinforcement improves the behavior of soils B and C. Soil A is again the soil that presents the worst behavior over all. Figure 2 shows the comparison of the results between reinforced and non-reinforced soil, both on a weak subgrade. Soils B and C are the soils where the reinforcement was effective in reducing the rut deformations, as the curves that

Table 5. Influence of the reinforcement on rut deformation

Soil	Weak subgrade		Soil with reinforcement		Deformation variation (%)
	Deformation (mm)	Number of cycles	Deformation (mm)	Number of cycles	
A	11.51	72	20*	75	+66.8**
B	12.04	3000	5.96	3000	−50.5
C	8.69	3000	6.73	3000	−22.6
D	3.7	3000	4.35	3000	+17.6

* Test stopped at the end of 72 cycles as the safety limit value for deformation was reached.
** Variation calculated for 72 cycles and 19.2 mm of deformation for reinforced soil.

correspond to the reinforced soil are always below the curves that correspond to the unreinforced soil. This is clear also from Table 5.

3.3   Influence of the Type of Soil

On analyzing the results, it seems evident that soil A is the most sensitive to the subgrade strength and does not benefit from the reinforcement. The data in Fig. 3 clearly shows that soil D undergoes the least deformation, is the least influenced by the strength of the subgrade and it seems to gain no benefit from reinforcement. In fact, it seems that reinforcement is counterproductive to improving the behavior of the road.

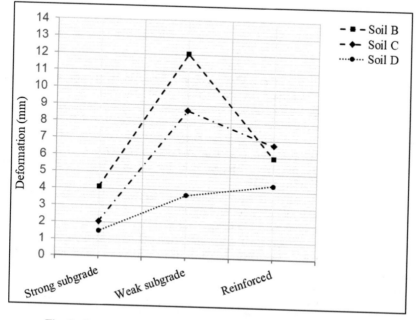

Fig. 3. Rut deformation at the end of 3000 cycles for all conditions

Soils B and C, the soils with intermediate granulometry were the most affected by the subgrade strength and they gained considerable benefit from the inclusion of the reinforcement. These effects are more pronounced in soil B.

4 Conclusions

From the study described herein, it can be concluded that:

- the strength of the subgrade is a very important factor, as lower strength means greater deformation, for all soil types;
- the inclusion of a reinforcement layer on top of the weak subgrade does not improve the performance of all soil types; soils A and D are negatively influenced by the inclusion of the reinforcement while soils B and C are positively influenced;
- soil A, the soil with the lowest fine particle content, performed worst of all, for all conditions studied, while soil D, which had the highest fine particle content, performed best of all, for all conditions studied;
- there is good evidence that the Wheel Tracking Machine is suitable for simulating traffic cycling loads on unpaved roads, and is therefore suitable for performing this type of study.

Acknowledgments. The authors thank the company BBF - Tecnologias do Ambiente (Environmental Technologies) for the supply of the geogrid and to ACIV for the financial support to present this paper at the GeoMEast 2019 International Congress & Exhibition.

References

ASTM 2487: Standard practice for classification of soils for engineering purposes (Unified Soil Classification System). American Society for Testing and Materials (2009)

ASTM 3282: Standard practice for classification of soils and soil-aggregate mixtures for highway construction purposes. American Society for Testing and Materials (2009)

Calvarano, L., Leonardi, G., Palamara, R.: Finite element modelling of unpaved road reinforced with geosynthetics. Transp. Geotech. Geoecol. **189**, 99–104 (2017)

EN 12697-22: Bituminous mixtures - Test methods for hot mix asphalt - Part 22: Wheel Tracking, European Standard (2007)

IGS: Geosynthetics in Unpaved Roads, International Geosynthetic Society (2018). www.geosyntheticssociety.org. Last accessed on July 2019

Reilly, C., Nell, K.: Geosynthetic reinforcement for unpaved roads - recent experience. In: Proceedings of the Conference: Civil Engineering Research in Ireland, pp. 252–257 (2018)

Tang, X., Abu-Farsakh, M., Hanandeh, S., Chen, Q.: Performance of reinforced-stabilized unpaved test sections built over native soft soil under full-scale moving wheel loads. J. Transp. Res. Board **2511**, 81–89 (2015)

Wu, H., Huang, B., Shu, X., Zhao, S.: Evaluation of geogrid reinforcement effects on unbound granular pavement base courses using loaded wheel tester. Geotext. Geomembr. **43**, 462–469 (2015)

Pavement Performance Evaluation and Maintenance Decision-Making in Rwanda

Li Bo, Marie Judith Kundwa$^{(\boxtimes)}$, Cui Yu Jiao, and Zhu Xu Wei

Lanzhou Jiaotong University, Lanzhou, China
kudith04@yahoo.fr

Abstract. Rwanda is one most developing country of East Africa Countries (EAC), every year has a great development in different areas. Regarding to infrastructure sector there is a great change in terms of housing, traffic (vehicles, trucks), etc., and when every day traffic loads and pavement age both increasing, it will gradually deteriorate and decrease functional and structural performance of a pavement. Deterioration of pavement can be attributed to various factors like age, traffic, environment, material properties, pavements thickness, strength of pavement as well as subgrade properties which affect the mechanical characteristics of a pavement. This research was conducted in order to assess the effect of Truck & Other heavy vehicles (CVPD), California Bearing Ratio (CBR), precipitation, pavement age and thickness factors on deflection and International Roughness Index (IRI) to find out which factors could be used in pavement performance evaluation in Rwanda as predictor variables and to assess the correlation between those variables. The result shows that precipitation and CBR found to be a significant predictor for both deflection and IRI on Rwandan flexible pavement performance and CBR is strongly correlated with precipitation. Therefore, the climate input precipitation was found to be more important factor for predicting different pavement performance in Rwanda, for further studies the temperature and Pavement Condition Index (PCI) need to be collected and analyzed, then the results would be compared to support the greater effectiveness of decision making and program development for Rwanda pavement performance evaluation.

Keywords: Pavement performance model · Deflection · Riding quality · Decision-making · Regression analysis

1 Introduction

Pavement performance is effectiveness or adaptability of road pavement conditions to meet different driving requirements, including functional performance and structural performance. Evaluation of existing flexible pavement condition is a requirement to choose improvement technique that has to be implemented to improve its quality [1]. Pavement management systems are a subset that have been in place for over 30 years, explicitly recognize the importance of maintenance and rehabilitation planning to ensure that the infrastructure assets remain viable [2]. These importance's include items such as ability to document the network condition, ability to predict future conditions given a variable budget, increased creditability among stakeholders [3]. The Pavement

© Springer Nature Switzerland AG 2020
S. Badawy and D.-H. Chen (Eds.): GeoMEast 2019, SUCI, pp. 107–116, 2020.
https://doi.org/10.1007/978-3-030-34196-1_8

Design and Management Guide developed by the Transportation Association of Canada [4] as well as the Pavement Management Guide developed by the American Association of State Highway and Transportation Officials (AASHTO) [5], provides useful information on pavement management processes including data requirements, data collection methods, pavement performance prediction, selection of maintenance and rehabilitation treatments, priority analysis, and other pavement management topics. Different types of data are required for managing the road infrastructure, Inventory data describe the physical elements of a road system, condition data describe the condition of elements that can be expected to change over time [6]. The sustainability of data collection is strongly influenced by the frequency of surveys. Data should be collected only as frequently as is required to ensure proper management of the road network. The frequency can vary depending upon the data of interest. The quality of data has to be taken in consideration because if the quality of data is not good, the preservation will not be cost-effective, to quantify the performance extension, quality of data is needed [7]. Pavement evaluations are performed in the field through manual surveys or using specialized equipment. The rate of pavement deterioration, the maintenance treatments appropriate for the pavement condition, the timing of the eventual rehabilitation, the costs incurred, are all part of a single performance period evaluation [8].

2 Pavement Performance Model

The ultimate criterion used for assessing pavement quality is pavement performance, in terms of indicators such as pavement distress, roughness, rutting, anti-skid resistance, structure strength and etc. [9]. In America, For Ontario highways, several Key Performance Index (KPIs) are used for Pavement management decisions, such a as PCI, Distress Manifestation Index (DMI), IRI, Riding Comfort Index (RCI) etc. [10] estimating the prediction models by using Ordinary Least Square (OLS) approach; estimated KPI models they considered all available effecting pavement performance by using 'Seemingly Unrelated Regression (SUR)' method and found that PCI model is highly correlated with DMI and RCI model. However, IRI model is not found highly correlated to other models. For South Carolina state, they developed a performance evaluation models using regression techniques for one of the Mechanistic Empirical Pavement Design Guide (MEPDG) performance indicators: IRI, and three of the South Carolina Department of Transportation (SCDOTS) pavement performance indices: PSI, PDI and PQI. Precipitation was found to be a significant predictor for PSI on both types of pavement Asphalt Concrete (AC) and Jointed Plain Concrete Pavements (JPCP) [11]. For Illinois Department of Transportation has performance models, they used deduct values based on historical performance, route type, pavement type, age, and presence of particular distresses [3]. In India, statistical and Artificial Neural Networking (ANN) modeling was done for performance prediction of low volume roads, it is observed that if pavements are classified on the basis of parameters influencing pavement condition such as pavement age, traffic, CBR of subgrade and pavement thickness then it predicts the pavement condition in a better way. Therefore Maintenance Priority index was developed using three parameters named as deflection, riding quality and traffic to decide the priority [12]. In Sudan, they applied PCI methodology in pavement distress evaluation and maintenance prioritization [13]. In china, based on the "Highway Performance Assessment Standard", the decision of pavement maintenance plan is determined by the MPI (Maintenance Plan Index), the sub-index of PQI and traffic volume [14]. The performance evaluation of pavement can cover many aspects including assessment of traffic safety on road, evaluation of road surface condition, structural adequacy of pavement and rideability of pavement surface. [15] have developed correlation analysis between pavement distresses and international roughness index by neural network, in that study, the coefficient between pavement distresses and IRI reaches 0.944. It shows that International Roughness Index may totally reflect on pavement distress conditions. Thus, IRI was used as a pavement performance index. [16] has developed pavement roughness and pavement condition model for Saudi highways. The pavement condition rating was calculated based on rutting, raveling, cracking and International Roughness Index They found the R^2 value 83.9% and 95% for Pavement Condition Rating and International Roughness Index by regression model.

3 Decision-Making Framework for Pavement Preservation

A management framework, when looking at roads, can therefore assist in improving the quality of decision-making, and can result in greater effectiveness and efficiency for both customers of the road network and the road administration [17]. In Rwanda, the emphasis is placed on the development of transport infrastructure and services, in terms of construction, rehabilitation and maintenance of the transportation networks [18] but it is a time to think about pavement management systems inventory historical and current conditions of roadway networks in order to predict the future conditions of such networks, and suggest schedules for maintenance, repair, and rehabilitation activities [19]. The Road Maintenance Fund (RMF) is an institution established by the law to ensure collection and funding for the maintenance of road networks in Rwanda. Unfortunately, these achievements are at a modest level in comparison to general maintenance needs throughout the country. This modest level is the consequence of insufficiency of funds [20]. Decision making for pavement maintenance and rehabilitation should be integrated into a yearly management cycle of planning, budgeting, engineering, and implementation activities. There are eight basic steps in the yearly management cycle: review or establishment of levels of service, pavement inventory, identification of needs, prioritization, budgeting, project design, project implementation, and performance monitoring [2].

4 Methodology

The main objectives of this research are to develop a pavement performance evaluation models in Rwanda by using multiple linear regression analysis.

5 Multiple Regression Analysis

Multiple regression analysis is almost the same as simple linear regression. The only difference between simple linear regression and multiple regression is in the number of predictors ("x" variables) used in the regression. The general purpose of multiple regression (the term was first used by Pearson, [21] is to learn more about the relationship between several independent or predictor variables and a dependent or criterion variable.

The general linear regression model is given by

$$Y_i = \varepsilon_0 + \varepsilon_1 X_{i1} + \varepsilon_2 X_{i2} + \varepsilon_{p-1} X_{i(p-1)} + \ldots \varepsilon_i, \quad i = 1, 2, \ldots n.$$

for this research:

Xi = Predictors variables (CVPD, precipitation of subgrade, thickness and ages of pavement)

Those factors were chosen according to data available given by Rwanda Transportation Department Agency (RTDA)

Y = Response variables (IRI and deflection)

6 Database

The network selected for the present study is the National Road number one (NR1) in Rwanda from Kigali to Akanyaru (Burundi border). This roadway network has been taken into 16 pavement sections sample with a length of 127.03 km. The NR1 has been designed to be upgraded to surfaced road standards in 1973–77 and the construction works have been executed in 1978–83. The pavement has been rehabilitated in 1998–2000 (Kigali – Muhanga) and in 2004–2005 (Muhanga – Akanyaru).

A geotechnical survey of the existing pavement with sampling and testing has been carried out by Consultants of RTDA and the necessary information has been acquired by destructive and non-destructive testing. The structural types of distress progressively affect the pavement's ability to support traffic loads are found like potholes, longitudinal cracking, fatigue alligator cracking, rutting. The deflection values have been statistically processed following the equivalent surfaces procedure recommended by AASHTO to determine the boundaries of the homogeneous sections. The existing Nyabarongo bridge is in good structural condition but it has been overtopped three times in the recent years (26/04/1998, 04/05/2002 and 19/04/2013) [22]. Referring to the results coming from the geotechnical survey, from the structural point of view, some existing pavement appears in fair structural condition others in good condition as shown in Table 1 with the data of the variables that can affect the road condition available in RTDA database (CVPD, precipitation of subgrade, thickness, age) detected in the year 2016 shown in Table 2.

7 Application and Discussion of Results

The following Table 1 is describing the identification of the selected road with its section. It gives us the information about the location of the road according to the district, the length and AADT.

Table 1.

Region	District	Road name	Starting point	Ending point	Road length (km)	AADT	AADT range
KIGALI CITY	Nyarugenge	RN1	0 + 000	1 + 950	1.95	9739	>5000 < 10000
KIGALI CITY	Nyarugenge	RN1	1 + 950	2 + 500	0.55	9739	>5000 < 10000
KIGALI CITY	Kicukiro	RN1	2 + 500	2 + 700	0.20	15202	>15000 < 20000
SOUTHERN	Kamony	RN1	2 + 700	5 + 000	2.30	15202	>15000 < 20000
SOUTHERN	Kamony	RN1	5 + 000	14 + 225	9.23	9987	>5000 < 10000
SOUTHERN	Kamony	RN1	14 + 225	14 + 225	0.00	9987	>5000 < 10000
SOUTHERN	Kamony	RN1	14 + 225	19 + 600	5.38	4771	>4500 < 5000
SOUTHERN	Kamony	RN1	19 + 600	30 + 400	10.80	4771	>4500 < 5000
SOUTHERN	Muhanga	RN1	30 + 400	41 + 900	11.50	4771	>4500 < 5000
SOUTHERN	Muhanga	RN1	41 + 900	68 + 500	26.60	4771	>4500 < 5000
SOUTHERN	Ruhango	RN1	68 + 500	74 + 500	6.00	1954	>1500 < 2000
SOUTHERN	Nyanza	RN1	74 + 500	90 + 000	15.50	1954	>1500 < 2000
SOUTHERN	Huye	RN1	90 + 000	100 + 200	10.20	1954	>1500 < 2000
SOUTHERN	Huye	RN1	100 + 200	107 + 500	7.30	1954	>1500 < 2000
SOUTHERN	Huye	RN1	107 + 500	125 + 025	17.53	1954	>1500 < 2000
SOUTHERN	Nyaruguru	RN1	125 + 025	127 + 029	2.00	1954	>1500 < 2000

The Table 2 contain the pavement condition data of the selected road as IRI is used by highway professionals throughout the world as a standard to quantify road surface roughness. Pavement roughness is defined as an expression of irregularities in the pavement surface that adversely affect the ride quality of a vehicle. Roughness is an important pavement characteristic because it does not only affect ride quality but also affects fuel consumption, vehicle delay costs and maintenance costs. Pavement surface deflection measurements are the primary means of evaluating a flexible pavement structure and rigid pavement load transfer. surface deflection is an important pavement evaluation method. Deflection is a function of traffic (type and volume), pavement structural section, temperature affecting the pavement structure and moisture affecting the pavement structure. Deflection measurements using back calculation methods to determine pavement structural layer stiffness and the subgrade resilience modulus.

The correlation between IRI and predictor variables are found as low to large, with the Pearson correlation values (r) ranging from 0.01 to 0.582. Precipitation is the strongest related predictor of IRI (r = 0.582, p < 0.01). The table also shows that some of the predictor variables have strong correlations with each other. For instance, CBR is strongly correlated with precipitation (r = 0.514, p < 0.01). In contrast, the correlation

Table 2.

Starting point	Ending point	% truck & Other heavy vehicles (CVPD)	IRI m/km	Deflection mm	Pavement thickness cm	CBR	Precipitation	Pavement age
0 + 000	1 + 950	1.42	3	0.6	4.5	28	24	17
1 + 950	2 + 500	1.42	6	0.6	4	24	24	17
2 + 500	2 + 700	21.06	6	0.6	3	24	24	17
2 + 700	5 + 000	21.06	6	0.6	3.4	30	46.5	17
5 + 000	14 + 225	17.9	6	1.1	4.6	52	93.2	17
14 + 225	14 + 225	17.9	6	1.05	5.3	52	56.8	17
14 + 225	19 + 600	7.82	6	0.67	5	30	79.8	17
19 + 600	30 + 400	7.82	6	1.14	4.25	44	98.3	17
30 + 400	41 + 900	7.82	8	1.28	4.3	52	107.3	12
41 + 900	68 + 500	7.82	8	1.88	4.8	52	106	12
68 + 500	74 + 500	13.31	6	1.38	5	44	101.4	12
74 + 500	90 + 000	13.31	8	1.68	3.5	44	105.2	12
90 + 000	100 + 200	13.31	6	0.54	5	28	106	12
100 + 200	107 + 500	13.31	6	1.04	4.7	52	94.5	12
107 + 500	125 + 025	13.31	6	0.89	4	28	101.6	12
125 + 025	127 + 029	13.31	6	0.59	4.9	24	92.6	12

between CVPD with age and CBR are found as low. Lastly, pavement age is highly correlated with IRI but negatively, means that with increasing of pavement age IRI decreases as shown in Table 3.

Table 3.

	IRI	Precipitation	Age	CBR	CVPD	Thickness
IRI	1					
Precipitation	.582	1				
Ages	−.498	−.737	1			
CBR	.473	.514	−.218	1		
CVPD	.168	.061	.010	.095	1	
Thickness	−.127	.384	−.210	.336	−.241	1

Table 4 is The Model Summary offers the multiple r and coefficient of determination (r^2) for the regression model. As can see $r^2 = .789$ which indicates that 78.9 of the variances in deflection can be explained by our regression model. In other words, the effect of Predictors (CVPD, ages, CBR, thickness, precipitation) are strongly related to deflection.

Table 4.

Model	R	R square	Adjusted R square	Std. error of the estimate
1	.888[a]	.789	.684	.23522

[a]Predictors: (Constant), CVPD, Ages, CBR, Thickness, Precipitation
[b]Dependent Variable: Deflection

By seeing ANOVA Table for verification if the model fit, the model explains a statistically significant proportion of the variance as represented by Table 5.

Table 5.

Model	Sum of squares	df	Mean square	F	Sig.	
1	Regression	2.073	5	.415	7.492	.004[b]
	Residual	.553	10	.055		
	Total	2.626	15			

[a]Dependent Variable: Deflection
[b]Predictors: (Constant), CVPD, Ages, CBR, Thickness, Precipitation

For deflection model developed for AC, CVPD, ages, CBR, thickness, precipitation showed statistically significant effects on deflection ($p < 0.01$)

Model coefficients shown by Table 6, gives the constant or intercept term and the regression coefficients (**b**) for each explanatory variable the constant value is (1.359). CBR and precipitation have positive effects on deflection, whereas thickness, age and CVPD has negative effects on deflection for flexible pavement in Rwanda. That means deflection increases with increasing CBR and increasing precipitation. However, deflection decreases with increasing thickness, pavement ages and CVPD. This indicates that for every unit increase in CVPD the model predicts a decrease of −0.014 in deflection, means that this decrease is not significant as showed by significance of 0.213 which is not less than 0.05, the same for ages and thickness of Pavement.

Table 6.

Model		Unstandardized coefficients		Standardized coefficients	t	Sig.
		B	Std. error	Beta		
1	(Constant)	1.359	.772		1.761	.109
	CBR	.028	.006	.785	4.382	.001
	Thickness	−.199	.107	−.314	−1.867	.092
	Precipitation	.002	.003	.151	.573	.579
	Ages	−.038	.036	−.232	−1.031	.327
	CVPD	−.014	.011	−.205	−1.332	.213

[a]Dependent Variable: Deflection

Table 7.

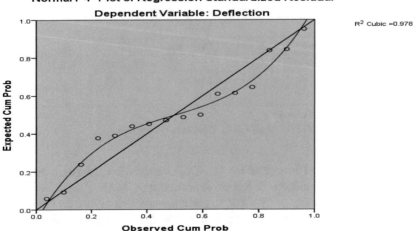

The Table 7 of P-P plot compares the observed cumulative distribution function (CDF) of the standardized residual to the expected CDF of the normal distribution. Note that are testing the normality of the residuals and not predictors.

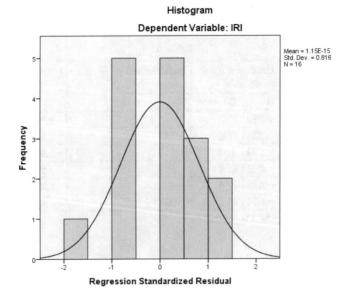

A Histogram of the residuals suggests that they are close to being normally distributed and there are more residuals close to zero than perhaps would expect.

8 Conclusion

Precipitation and CBR found to be significant predictor for both deflection and IRI on Rwandan flexible pavement ($p < 0.05$). Therefore, the climate input precipitation was found to be more important for predicting different pavement performance in Rwanda, for further studies the temperature need to be collected and the results would be compared. Pavement age and its thickness were found to be highly correlated with IRI but negatively, means that with increasing of pavement age, IRI decreases. The effect of considered predictors for the research (CVPD, ages, CBR, thickness, precipitation) are strongly related to deflection more than to IRI. CBR of soil subgrade have statistically significant effects on Rwandan pavement performance. But there is a need of more different types of data for managing the road infrastructure, inventory data that describe the physical elements of a road system and condition data that describe the condition of elements in Rwandan pavement management database and those kinds of data must be objective, reliable, useful and repeatable. Decision-Making for pavement maintenance and rehabilitation in Rwanda have to be integrated into a yearly management cycle of planning, budgeting, engineering, and implementation activities in order to assist the quality of Decision-Making for Rwandan pavement performance.

References

1. Aghera, H.V., Mandhani, J., Solanki, R.V.: A review on performance evaluation of flexible pavement, s.l. Int. J. Adv. Eng. Res. Dev. **4**(2), 2348–4470 (2017)
2. Hajek, J., Hein, D., Olidis, C.: Decision making for maintenance and rehabilitation of municipal pavements, s.n. In: Annual Conference of the Transportation Association of Canada, Québec City, Québec (2004)
3. Gary Hicks, R., Simpson, A.L.: Pavement Management Practices in State Highway Agencies: Madison, Wisconsin Peer Exchange Results. Federal Highway Administration US Department of Transportation, Madison, Wisconsin. FHWA-HIF-11-035 (2011)
4. TAC: Pavement Design and Management Guide, s.n., 2323 St. Laurent Boulevard, Ottawa, K1G 4J8 (1997)
5. AASHTO: Pavement Management Guide. AASHTO Joint Task Force on Pavements. American Association of State Highway and Transportation Officials (2001). ISBN-1-56051-155-9
6. Bennett, C.R.: Data collection technologies for pavement management systems, s.n. In: 7th International Conference on Managing Pavement Assets, 1818 H Street NW, Washington, DC 20433 (2008)
7. Geoff Hall, P.E.: Perspective on pavement condition data for pavement preservation, s.n. In: National Pavement Preservation Conference, Maryland SHA (2016)
8. Luhr, D.R., Rydholm, T.C.: Economic evaluation of pavement management decisions. In: 9th International Conference on Managing Pavement Assets, Alexandria (2015)
9. Li, Q., Liu, G., Pan, Y.: Study on life-cycle cost analysis based pavement maintenance design methodology, s.l. In: 8th International Conference on Managing Pavement Assets, ICMPA096 (2007)
10. Jannat, G.E., Tighe, S.L.: Performance based evaluation of overall pavement condition indices for Ontario highway systems, s.n. Session of the 2015 Conference of the Transportation Association of Canada, Charlottetown, Canada (2015)

11. Rahman, M.M., Uddin, M.M., Gassman, S.L.: Pavement performance evaluation models for South Carolina. KSCE J. Civ. Eng. **21**, 2695–2706 (2017)
12. Gupta, Ankit, Kumar, Praveen, Rastogi, Rajat: Pavement deterioration and maintenance model for low volume roads. Int. J. Pavement Res. Technol. **4**, 195 (2011)
13. Mergi, K.M., Mohamed, E.K.M.: Application of Pavement Condition Index (PCI) methodology in pavement distress evaluation and maintenance prioritization, s.n. In: Annual Conference of Postgraduate Studies and Scientific Research (Basic and Engineering studies Board), Friendship Hall, Khartoum (2012)
14. Gao, S., Wu, P., Feng, J.: Pavement, decision-making method of maintenance scheme for highway asphalt. Trans Tech Publications, Switzerland (2013)
15. Lin, J., Yau, J., Hsiao, L.: Correlation analysis between International Roughness Index (IRI) and pavement distress by neural network, s.l. Transportation Research Board, pp. 1–21 (2003)
16. Mubaraki, M.: Development of pavement condition rating model and pavement roughness model for saudi highways, s.l. Adv. Mater. Res. **723**, 820–828 (2013)
17. Kerali, H. R., Robinson, R: The Role Of HDM-4 in Road Management, s.n., South Africa (2000)
18. Lombard, P.: Rwanda Strategic Transport Master Plan. Rwanda Transport Development Agency, KIgali (2012)
19. Kuhn, K.: Pavement Network Maintenance Optimization Considering Multidimensional Condition Data. D. 4, s.l., vol. 18 American Society of Civil Engineers (2012)
20. RMF Annual Progress Report. Mukamurenzi. Nociata. Road Maintenance Fund, Kigali (2015)
21. Nathans, L.L., Oswald, F.L., Nimon, K.: Interpreting multiple linear regression, s.l. Pract. Assess. Res. Eval. **17**(9) (2012)
22. S.P.A, AIC PROGETTI: Consultancy services for the study of Kigali – Muhanga – Huye – Akanyaru road (157 km). CONTRACT N° 019/RTDA/015, Kigali (2016)

ANN-Based Fatigue and Rutting Prediction Models versus Regression-Based Models for Flexible Pavements

Mostafa M. Radwan[1]([✉]), Mostafa A. Abo-Hashema[2],
Hamdy P. Faheem[3], and Mostafa D. Hashem[3]

[1] Department of Civil Engineering, Nahda University, Beni Suif, Egypt
`mostafa.yaseen@nub.edu.eg`
[2] Department of Civil Engineering, Fayoum University, Fayoum, Egypt
[3] Department of Civil Engineering, Minia University, Minia, Egypt

Abstract. Roads are exposed to continuous deterioration because of many factors such as traffic loads, climate and material characteristics. In Middle East countries, incredible investments have been made in constructing roads that necessitate conducting periodic evaluation and timely maintenance and rehabilitation (M&R) plan to keep the network operating under acceptable level of service. The M&R plan necessitates performance prediction models, which represent a key element in predicting pavement performance. Consequently, there is always a need to develop and update pavement performance prediction models specially for fatigue and rutting distresses, which are considered the most major distresses in asphalt pavement. On the other hand, Artificial Neural Network (ANN) is considered the best solution to developing such models with high accuracy due to its brilliant mechanism in training, testing and evaluating the data. In addition, the ANN approach has the flexibility to change many parameters such as number of neurons, hidden layers and function type to obtain more accurate predicted models. The scope of this paper is to develop ANN-based fatigue and rutting prediction models for asphalt roads. The ANN-based models were developed using MATLAB 2017b software based on actual field data obtained from Long-Term Pavement Performance (LTPP) database. The models were developed for both wet and dry non-freeze climatic zones. Results indicated that the ANN approach can be used in predicting both fatigue and rutting distresses with high accuracy as compared with the developed statistical models' approach, which were also developed in this study for both fatigue and rutting distresses.

Keywords: Prediction models · Distress models · LTPP · Neural network · Modelling · Climatic zone · Maintenance activities · ANN

1 Introduction

The Middle East countries are experiencing tremendous growth in infrastructure especially in constructing asphalt roads, which require periodic Maintenance and Rehabilitation (M&R) activities to preserve such investments. To identify M&R

© Springer Nature Switzerland AG 2020
S. Badawy and D.-H. Chen (Eds.): GeoMEast 2019, SUCI, pp. 117–133, 2020.
https://doi.org/10.1007/978-3-030-34196-1_9

activities based on yearly basis, future pavement condition should be predicted using pavement performance prediction models. This task should be implemented through application of well-designed Pavement Management System (PMS) (Haas and Zaniewski 1994; Hajek 2011; Zimmerman and Testa 2008). One of the main purposes of the PMS is to come up with the most cost-effectiveness policies for M&R activities. Pavement performance is predicted using distress/performance prediction models, which are considered the heart of the PMS system to quantify pavement deterioration rate and hence identifying M&R activities in a timely M&R plan supported by budget requirements (Zimmerman and Testa 2008; Naiel 2010).

Fatigue cracking and rutting or permanent deformation are considered two major distresses in asphalt pavements that cause structural failure in pavement layers. Horizontal tensile strain at the bottom of the asphalt layer(s) is the main cause for fatigue cracking; whereas rutting is produced due to vertical compressive strain on the top of subgrade layer due to weakness of foundation.

Consequently, there is always a need to develop and to update pavement performance prediction models embedded in PMS applications. Various performance prediction models had been introduced through the years, some of which are considered simple, while others are quite complex. There are two streams of pavement performance modelling, which are deterministic and stochastic approaches. The major differences between deterministic and stochastic performance prediction models are model development concepts, modelling processor formulation, and output format of the models. There are different types of deterministic models, such as mechanistic models, mechanistic-empirical models, and regression or empirical models. The mechanistic models draw the relationship between response parameters such as stress, strain, and deflection. The mechanistic-empirical models are oftentimes established in association with design systems and hence have not been broadly used in PMS but rather can possibly be applied at a network level. On the other hand, the regression models represent the link between the performance parameters such as pavement distresses and the forecasting parameters such as age, traffic loading, pavement material properties, and thickness (Abo-Hashema 2013; Mubaraki 2010; Radwan et al. 2019).

On the other hand, Artificial Neural Network have been effectively utilized for some errands including pattern recognition, optimization, function approximation, data retrieval, and predicting (Mubaraki 2010). ANN utilizes the mathematical emulation of genetic nervous systems in order to process acquired data and deduce predictive outputs after training the network suitably for pattern recognition (Thube 2012). A neural network comprises of various layers of parallel preparing component, or neurons. At least one hidden layer exists among an input layer and output one. The hidden layers neurons are associated with the neurons of a neighbouring layer through weighting factors which are adjustable via the training procedure of the model. The networks are structured according to training procedures for particular applications (Thube 2012).

This study focuses on developing ANN-based fatigue and rutting prediction models for asphalt roads located in both wet and dry non-freeze climatic zones, which represent most of the Middle East countries such as Egypt using data extracted from the Long-Term Pavement Performance (LTPP) database. Although, the ANN results were acceptable and more satisfied to predict Fatigue and Rutting distresses, the present study seeks for applying a comparison with another approach to show the powerful ability of ANN approach and to get the maximum benefit in forecasting fatigue and rutting distresses. Consequently, concerted efforts were also conducted in this study during another phase to develop regression prediction models through deterministic approach to predict fatigue and rutting distresses. Then, an interesting comparison was performed between results of the developed ANN-based fatigue and rutting models against similar developed models based on deterministic approach.

Based on the results, it was found that ANN-based models are appropriate to predict fatigue and rutting distresses with high accuracy due to its brilliant mechanism in testing and evaluating data. This study is considered as a crucial attempt to not only develop such models for the Middle East countries due to lack of resources led to unavailability of such models in most of Middle East countries, but also to compare between ANN-based and Regression-based fatigue and rutting prediction models.

2 Overview of Pavement Distress Models

Several pavement distress prediction models had been introduced through the years. The models differ significantly in their generality, their capability to forecast pavement performance with acceptable accuracy, and requirements of input data. Many of these models are empirical and were produced for use under specific traffic and climatic environments. Some of the models are mechanistic-empirical in which the input parameters are estimated by mechanistic models.

Performance prediction model is defined as a mathematical formula to predict future pavement deterioration depending on the current pavement condition and other affecting factors (Mubaraki 2010). Historical database for measures of pavement condition, age and traffic are extremely important in fitting forecast pavement deterioration models. These models are the major input to the effective PMS (Mubaraki 2010).

On the other hand, pavement distress prediction models are exceptionally powerful for basic leadership process in setting up answers to the inquiries of what, where, when, concerning support maintenance needs. Additionally, these models are crucial in identifying a timely M&R plan and when the action plan should start to keep the road network under acceptable level of service (Vepa et al. 1996).

The factors that could affect fatigue prediction models include age, traffic loading, pavement condition data, climatic condition, material characteristics, and quality of construction and maintenance. On the other hand, the factors that could affect rut depth distress prediction models include internal factors, such as asphalt binder, air voids in total mix, layers thickness, voids in the mineral aggregate, Marshall stiffness, subgrade material stiffness, elastic modulus of asphalt layer; and external factors such as traffic loading and environmental related factors. The availability and accuracy of data definitely affect the confidence level of the prediction model.

3   Background of LTPP

The Long-Term Pavement Performance program (LTPP) is the largest pavement performance research program ever undertaken, gathering data from more than 2,000 pavement test sections over a 20-year test period. The single most significant product of the LTPP program is the pavement database - the largest and most comprehensive collection of research-quality performance data on in-service highway pavements ever assembled. LTPP is one of the significant research regions of the Strategic Highway Research Program (SHRP). Strategic Highway Research Program was the supportive for LTPP program for the first initial five years. The Federal Highway Administration (FHWA) had proceeded along with the administration and subsidizing of the program, since 1991. The LTPP program was overseen via the LTPP Team under the Office of Infrastructure Research and Development (Radwan et al. 2019; Abo-Hashema and Sharaf 2009; LTPP 2017; FHWA 2002).

There are two complementary experiments inside LTPP to achieve the objectives. First, the General Pavement Studies (GPS) utilize the originally constructed current pavements after the initial overlay and concentrate on the most frequently used pavement structural design. Specific Pavement Studies (SPS) is considered the second series of LTPP experiments whose test sections let the factors of critical design to be performed, controlled, and monitored from the construction date. The results will offer a preferable understanding of the way to select M&R and design factors which influence pavement performance. GPS and SPS sets comprise of more than 2,500 test segments situated on all through North America built in four climate zones: wet-non-freeze, wet-freeze, dry-non-freeze, and dry-freeze. The LTPP program screens and gathers asphalt execution information on every single dynamic site. The gathered information incorporates data to develop seven modules which are: Maintenance, Inventory, Rehabilitation, Monitoring (Distress, Deflection, and Profile), Traffic, Materials Testing, and Climatic. The LTPP Information Management System (IMS) is considered the focal database whereas the information gathered by the program of LTPP. This database is persistently being produced as more information is gathered and handled (Abo-Hashema 2013; Radwan et al. 2019).

4   Artificial Neural Networks

ANNs are later computational models characterized in similarity with the natural attributes to recreate the choice procedure in the cerebrum. They are helpful to inexact and estimate unknown functions relying upon different and various input esteems. One of the principle attributes of this methodology is that it speaks to an approach to solve very complicated and nonlinear issues utilizing only very modest mathematical process. Specifically, ANN can be considered as a "black-box" method, since the outcomes are created without any respects to the causal connections among input and output. The strategy probability is completely misused when embraced for big data analysis and it very well may be utilized to create generalized solutions for issues utilizing big series of data. Like the cerebrum, the ANN is comprised of different interconnected neurons, which get input, process the data, and produce output for other connected neurons (Sollazzo et al. 2017).

In this study, MATLAB software, version 2014b, was used as a tool in developing a neural network. A multilayer feed-forward backprop ANN model is considered the most widely used neural network. The system incorporates input layer, at least one hidden layer and the output layer. Each artificial neuron gets and process data entering from different neurons and after that hand-off the signs to other neurons. Figure 1 shows Typical structures of ANN (Sollazzo et al. 2017; Abo-Hashema 2013).

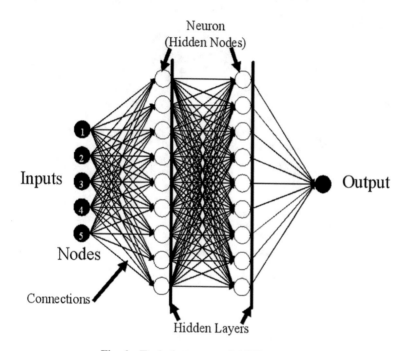

Fig. 1. Typical structure of ANN approach

A lot of papers discussed many applications of ANN on pavement Engineering such as developed an ANN for pavement condition evaluation, predicting present serviceability index (PSI), forecasting pavement performance using International Roughness Index (IRI), deducing of cracking progression, and studying of the variables affecting the compaction stage.

5 ANN-Based Fatigue and Rutting Prediction Models

5.1 Methodology

Figure 2 depicts the methodology adopted in developing the required distress prediction models for fatigue and rutting distresses based on ANN approach using MATLAB software version 2014b. The first crucial step is to create a database including all required data related to the required distresses. The database was created using LTPP sites located in wet and dry non-freeze climatic zones. Two database sets were created, one for training procedure and other for testing procedure. Sensitivity analysis using different numbers of hidden nodes and layers was also conducted on the trained ANN.

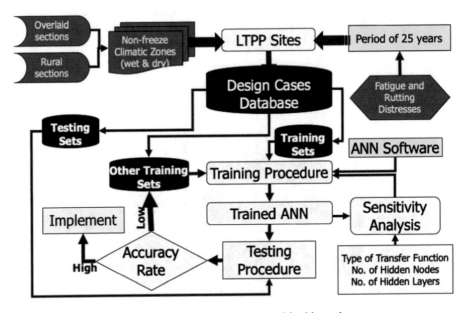

Fig. 2. Methodology implemented in this study

5.2   Design Cases Database

LTPP was the main source of data. Therefore, LTPP sites were selected to obtain the required data according to specific criteria as follows:

- Sites located in wet and dry non-Freeze climatic zones
- Only overlaid sections were chosen to simulate newly constructed pavement.
- Rural sections were selected represented main roads.
- Design period or data range was selected for 25 years, starting from 1991

Accordingly, 43 and 57 LTPP sites were selected for wet and dry non-freeze climatic zones, respectively. Data collection step was then started for the following data that related to Fatigue and Rutting distresses:

- Air temperature (Ta)
- Pavement age since overlay (PA)
- Traffic loading represented by Equivalent Single Axle Load (ESAL)
- Annual Precipitation
- Available pavement distresses
- Asphalt pavement thickness (T)

- Material characteristics:
  - Resilient modulus of subgrade soil (Mr)
  - % Passing the #200 sieve (0.075 mm) of subgrade soil (P_{200}),
  - % Air voids of asphalt mix (V_a),
  - % asphalt content in the mix (P_b)
  - Moisture content of base/subbase courses (MC_b),
  - Moisture content of subgrade soil (MC_S), and
  - Plasticity index of subgrade soil (PI)

All data were collected on different dates during the 25-year data range. The collected data have been filtered through a screening process to come up with feasible data that could be used to develop the required ANN models. The criteria for screening process are selected as follows:

1. Unavailability and/or insufficient of some distresses data
2. Absence of material characteristics data
3. Illogical data patterns, e.g. distress density should be increased with time not decreased

Consequently, 42 LTPP sites out of 43 were selected for wet non-freeze climatic zone; and 34 LTPP sites out of 57 were selected for dry non-freeze climatic zone, as shown in Table 1. The unit of distress data recorded in the LTPP database is based on the distress types. The unit of area is accounted for fatigue; on the other hand, the unit of length or depth is accounted for rutting distress. In addition to the collected distress data, distress density was calculated by dividing the length or area of distress by the area of examined section based on the PAVER system (Shahin and Kohn 1981).

For ANN-based fatigue and rutting prediction models, different inputs parameters are selected, and one output is required, which is fatigue distress density or rut depth. Sample of collected data is shown in Tables 2 and 3 for wet- and dry-non-freeze climatic zones, respectively (Radwan et al. 2019).

Two database sets were created, which are training and testing database. Other training database set was also created and could be used in case of low accuracy rate based on testing procedure.

All sets of training data were utilized to estimate error gradient and update weights and biases of the network. Moreover, the validation set error was monitored through the training procedure. In case of increasing the validation error, training had to be stopped.

5.3 Training Procedure

To develop performance prediction models for Fatigue and Rutting distresses based on ANN approach, the ANN network should be trained well using training database. The training database consists of 206 design cases for both fatigue and rutting distresses. The process was conducted using MATLAB software. It is noteworthy that MATLAB software divided the training dataset into two sets for training and validation.

Table 1. Selected non-freeze LTPP sites.

Site ID	State	Site ID	State
Wet-Non-Freeze Climatic Zone			
12-3997	Florida (FL)	28-2807	Mississippi (MS)
12-3996	*Florida (FL)*	28-3081	Mississippi (MS)
12-4106	Florida (FL)	37-1024	North Carolina (NC)
12-4107	Florida (FL)	37-1030	North Carolina (NC)
12-4108	*Florida (FL)*	37-1802	North Carolina (NC)
12-4097	Florida (FL)	40-1017	Oklahoma (OK)
12-9054	Florida (FL)	40-4163	Oklahoma (OK)
13-4096	Georgia (GA)	40-4087	Oklahoma (OK)
13-4112	Georgia (GA)	40-4161	Oklahoma (OK)
13-4113	Georgia (GA)	40-4165	Oklahoma (OK)
13-4111	*Georgia (GA)*	45-1025	South Carolina (SC)
13-4420	Georgia (GA)	5-3048	Arkansas
1-1021	Alabama (AL)	**48-3729**	**Texas (TX)**
1-4126	Alabama (AL)	**48-1113**	**Texas (TX)**
1-4129	Alabama (AL)	**48-1116**	**Texas (TX)**
1-1001	Alabama (AL)	**48-1093**	**Texas (TX)**
1-1019	Alabama (AL)	48-1068	Texas (TX)
24-1632	**Maryland (MD)**	48-1060	Texas (TX)
28-1001	Mississippi (MS)	48-3609	Texas (TX)
28-3028	Mississippi (MS)	**51-1023**	**Virginia (VA)**
28-3091	**Mississippi (MS)**	51-2021	Virginia (VA)
Dry-Non-Freeze Climatic Zone			
4-1002	Arizona (AZ)	**35-0108**	**New Mexico (NM)**
4-1003	Arizona (AZ)	35-0103	New Mexico (NM)
4-1006	Arizona (AZ)	**35-0104**	**New Mexico (NM)**
4-1007	Arizona (AZ)	35-0106	New Mexico (NM)
4-1015	Arizona (AZ)	35-0105	New Mexico (NM)
4-1017	Arizona (AZ)	**35-1112**	**New Mexico (NM)**
4-1021	Arizona (AZ)	35-0107	New Mexico (NM)
4-1024	Arizona (AZ)	35-0109	New Mexico (NM)
4-1025	Arizona (AZ)	35-0110	New Mexico (NM)
4-0113	**Arizona (AZ)**	35-0112	New Mexico (NM)
4-1062	Arizona (AZ)	**35-0101**	**New Mexico (NM)**
4-0160	Arizona (AZ)	48-1111	Texas (TX)
4-1065	Arizona (AZ)	48-1061	Texas (TX)
4-6055	Arizona (AZ)	48-1076	Texas (TX)
6-8151	California (CA)	48-3769	Texas (TX)
6-2004	California (CA)	48-6060	Texas (TX)
35-0101	New Mexico (NM)	**48-1048**	**Texas (TX)**
Sites to be selected for validation process			

Table 2. Sample of collected data for wet-non-freeze LTPP site.

%Density	Ta, °C	PA, Years	Mr, MPa	P_{200}	%Va	$\%MC_b$	$\%MC_S$	PI
Fatigue cracking model								
0	24.30	4	114	–	–	4	7	–
6.67	19.40	14	73	3.50	–	4	7	2
16.67	21.90	16.16	65	9.40	–	3	15	–

Rut depth mm	Ta, °C	PA, Years	ESAL	Annual precipitation	%Va
Rutting model					
6	15.89	5.92	711	1778.5	7.091
8	16.89	15.3	59	1679.30	5.823
10	15.60	12	40	1290.30	7.09
15	19.79	9.66	106	1418.59	3.993

Table 3. Sample of collected data for dry-non-freeze LTPP sites.

%Density	Ta, °C	PA, Years	Mr, MPa	P_{200}	%Va	T, mm	$\%MC_b$	$\%MC_S$	PI
Fatigue cracking model									
11.8	17.6	15.5	87	–	–	221	5	11	30
36.67	19.1	15.58	37	–	–	53.3	3	7	0
37.7	18.5	17.41	114	–	–	63.5	2	9	9

Rut depth mm	Ta, °C	PA, Years	ESAL	Annual precipitation	%Va
Rutting model					
11	22.70	18.25	925	121.9	16.3
7	23.10	16.58	768	41	16.3
5	17.70	15.5	4	343.4	6.12
4	16.10	8.416	12	294.6	6.12

5.4   Sensitivity Analysis

The aim of this step is to evaluate the fitness of the developed neural networks as an efficient way in predicting fatigue and rutting distresses with the most achievable accuracy that can be obtained. The neural networks are impressed by numerous parameters that can ensure the greatest possible accuracy such as transfer functions, number of nodes or neurons, and number of hidden layers.

A multilayer feed-forward backprop ANN model and TANSIG transfer function are developed to predict pavement fatigue and rutting distresses. The mean square error (MSE) and the coefficient of determination (R^2) were utilized to set the goodness or performance of models. The R^2 is defined as the proportion of the variance in the dependent variable that is predictable from the independent variable(s). A higher estimation of R^2 and lower MSE esteem guarantee a superior execution of the model and are increasingly valuable for forecast (Shafabakhsh et al. 2015).

The model precision of ANN relies upon the network architecture. Choosing quantity of the neurons in hidden layer doesn't have any broad guideline (Shafabakhsh et al. 2015). Distinctive ANN structures had attempted as far as cycles and hidden layer numbers. Rutting-Dry Neural Network model was considered as an example for showing graphs due to massive number of graphs in the study.

Figure 3 depicts the MSE estimations of networks against different neurons in hidden layer for Rutting-Dry Neural Network model. As shown in Fig. 3, the ANN with 8, 8, 18 and 20 neurons of hidden layer provided an impression of being the most ideal structure for predicting fatigue and rutting (Wet and Dry) distresses, respectively.

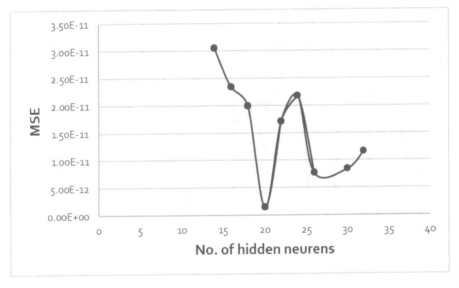

Fig. 3. Performance of rutting-dry ANN model under different No. of neurons

As shown in Fig. 3, Dry ANN model with 20 neurons has the lower MSE. Therefore, it is considered the most appropriate model for forecasting Rutting-Dry ANN model.

5.5   Testing Procedure

After the network was trained, testing procedure should begin using testing database extracted from the main developed database. The testing dataset have to be different from the data used in the training procedure.

The fitting graph between predicted and measured values utilizing the created Rutting-Dry Neural Network is shown in Fig. 4. As shown, the predicted values are close to measured values. This shows a solid relationship among the input parameters of the ANN model and the outputs.

As shown in Fig. 4, R^2 of data training, validation and testing values are 0.8183, 0.9199, and 0.8026, individually. Consequently, R^2 values achieved through ANN modelling method in the study are more than 0.8223 for all sets. The results revealed

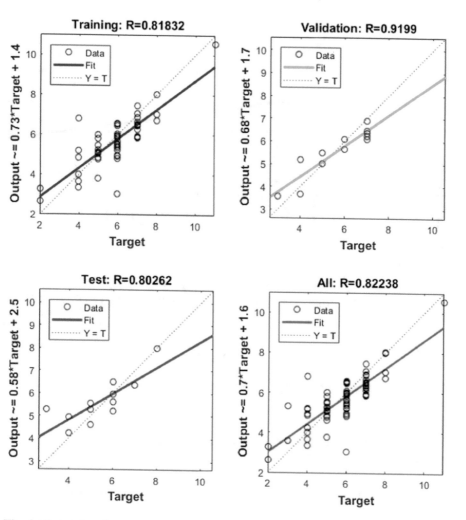

Fig. 4. Comparison between measured and predicted values for dry non-freeze rut depth by ANN for training data, validation data, testing data, and all data

that the established model has the capability to achieve at least 82% of the measured data for this model.

Figures 5 and 6 show Error in predicting rut depth for Dry Non-Freeze for training and testing Dataset, respectively. It can be seen from the figures that the predicted rut depth for Dry Non-Freeze accompanied with low errors, which are considered positive for forecasting rut depth values.

According to this, it tends to be reasoned that the suggested neural network can take in the connection among the distinctive input parameters and outputs. It creates the impression that established values from the ANN model considered genuinely near the actual values; also, they are equipped for propagating the input factors and outputs with high exactness of forecast.

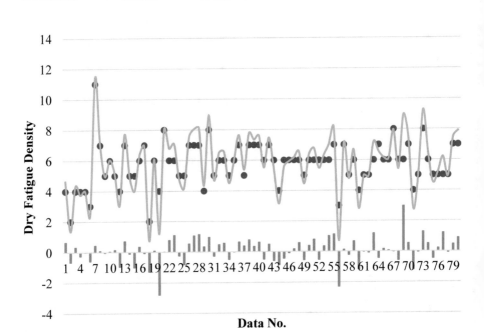

Fig. 5. Relationship between calculated and predicted rut depth in dry non-freeze zone for training set.

6   Development of Regression-Based Fatigue and Rutting Predictions Models

The main objective of the study is to develop pavement performance prediction models. The objective was achieved in many study phases. The first and second phases of this study were to develop regression-based pavement deterioration models for fatigue, rutting, bleeding, ravelling, longitudinal, and transverse distresses. The third phase of this study was to develop ANN-based fatigue and rutting prediction models and to compare with regression-based models, which is the subject of this paper. Other phases are related to comparison between developed models and available published models; in addition to implementation of the developed models in Egypt roads network, which necessitates performance data in a yearly basis.

A comparison between the ANN-based fatigue and rutting prediction models with similar ones developed by regression models to examine the suitability of using such models. Therefore, regression-based fatigue and rutting prediction models, which were developed during the first and second phases of this study, were used in this comparison (Radwan et al. 2019).

Stepwise regression test was performed within 95% confidence interval to come up with the most effective factors that could affect fatigue cracking and rutting distresses.

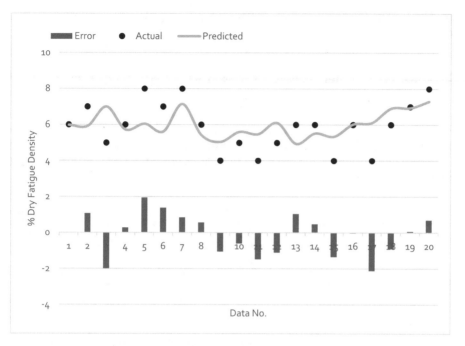

Fig. 6. Relationship between calculated and predicted for rut depth in dry non-freeze for testing set.

Hence, multiple regression analysis technique was applied to develop fatigue cracking and rut depth prediction models for wet and dry non-freeze climatic zones using SPSS software. Several trials were made to develop the required models that best represents the relation between the distresses with related factors. Therefore, the proposed distress models of fatigue cracking and rut Depth could be written as follows (Radwan et al. 2019):

Wet-non-freeze zone:

$$\%Fatigue\,Cracking \;=\; e^{(-10.356\,+\,1.936x\sqrt{PA}\,+\,1.422x\sqrt{MC_S})} \tag{1}$$

$$Rut\,Depth \;=\; 10.097 - 0.987\,x\,Ln(ESAL) + 0.478xV_a \tag{2}$$

Dry-non-freeze Zone:

$$\%Fatigue\,Cracking = e^{(-45.28\,+\,9.26\sqrt{PA}\,+\,2.1\sqrt{PI})} + 6.14\cos Ta \tag{3}$$

$$Rut\,depth = 21.39 + 0.009xESAL - 1.05 * Ta + 0.255xV_a \tag{4}$$

The study of statistical analysis approach showed that goodness of developed models was found to have bad fit with the same data trends with insufficient accuracy, which strengthen applying ANN approach in forecasting fatigue and rutting prediction models (Radwan et al. 2019).

7   Comparison Between ANN-Based and Regression-Based Prediction Models

Table 4 presents a comparison between developed ANN-based and Regression-based models for prediction of fatigue and rutting distresses. The comparison is made to show the fitness of each approach in the light of goodness of models which include R^2, MSE values and percent of error.

Table 4. Comparison between ANN and Regression Approaches for both R^2 and MSE values

Distress model	Climate	Statistical regression approach		ANN approach	
		R^2	MSE	R^2	MSE
Fatigue cracking	Wet	0.544	488.633	0.999	$1.23e^{-12}$
	Dry	0.465	482.729	0.937	$8.23e^{-13}$
Rut depth	Wet	0.233	693.294	0.977	$2.8e^{-11}$
	Dry	0.479	411.999	0.822	$1.42e^{-12}$

It was shown from Table 4 that R^2 value of most developed models for ANN approach is approximately twice of that R^2 value of the same developed models by statistical regression approach. Also, MSE of all developed models by ANN approach reached approximately to zero. Generally, R^2 value for all developed models by ANN approach is not less than 0.822, which confirms that ANN approach is the intelligent solution to predict fatigue and rutting models in both wet and dry non-freeze zones.

Figures 7, 8, 9 and 10 depict comparison between ANN-based and Regression-based prediction models through the difference in %error for predicted pavement distress fatigue and rutting (Wet and Dry), respectively. It was clear from graphs that the average of %error value for fatigue and rutting (Wet and Dry) by statistical approach is 4 times the value of the average %error for the same models by ANN approach, which strongly confirms that ANN approach is the magic technique for predicting fatigue and rutting (Wet and Dry) models.

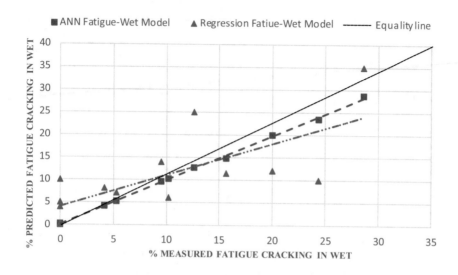

Fig. 7. Measured vs. predicted values for ANN-based and regression-based fatigue wet models

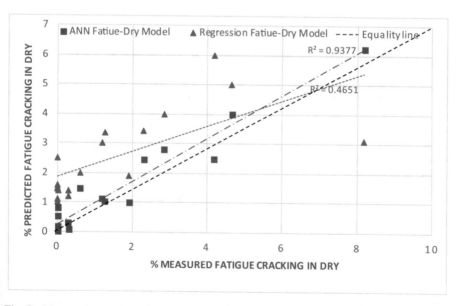

Fig. 8. Measured vs. predicted values for ANN-based and regression-based fatigue dry models

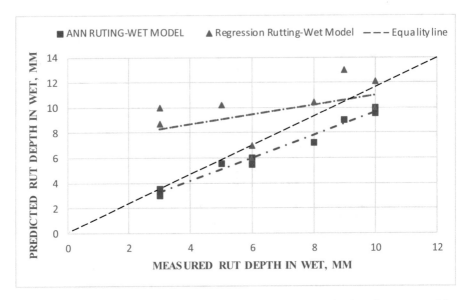

Fig. 9. Measured vs. predicted values for ANN-based and regression-based rut wet models

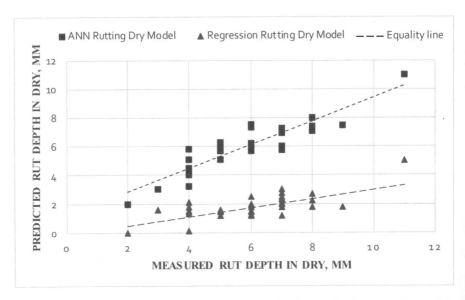

Fig. 10. Measured vs. predicted values for ANN-based and regression-based rut dry models

8   Conclusions

Two models were developed to forecast pavement fatigue and rutting distresses for both wet and dry non-freeze climatic zones using ANN approach. Furthermore, Regression-based models were also developed using the same data to predict fatigue and rutting prediction models. Moreover, a comparison between the two approaches was conducted. Based on the comparison and evidences of mean square error (MSE), the coefficient of determination (R^2), and %error values, results showed that the develoepd prediction models using ANN approach can be utilized in forecasting both fatigue and rutting distresses with high accuracy as compared to developed statistical models due to its brilliant mechanism in testing and evaluating a lot of data.

References

Long Term Pavement Performance (LTPP). https://infopave.fhwa.dot.gov/Data/DataSelection. Accessed 2017

Abo-Hashema, M.A.: Modeling pavement temperature prediction using artificial neural networks. In: Airfield and Highway Pavement Proceedings 2013 Airfield & Highway Pavement, USA, pp. 490–505 (2013). https://doi.org/10.2139/ssrn.3351973

Abo-Hashema, M.A., Sharaf, E.A.: Development of maintenance decision model for flexible pavements. Int. J. Pavement Eng. **10**(3), 173–187 (2009). https://doi.org/10.1080/10298430802169457

Federal Highway Administration (FHWA) (2002). https://www.fhwa.dot.gov/research/tfhrc/programs/infrastructure/pavements/ltpp/. Accessed August 2017

Haas, R.H., Zaniewski, W.J.: Modern Pavement Management System. Krieger Publishing Company, Malabar (1994)

Hajek, J.J.: Common airport pavement maintenance practices. Transportation Research Board (2011)

Mubaraki, M.: Predicting deterioration for the Saudi Arabia Urban road network. Ph.D thesis. University of Nottingham (2010)

Naiel, A.K.: Flexible pavement rut depth modeling for different climate zones (2010)

Radwan, M.M., Abo-Hashema, M.A., Faheem, H.P., Hashem, M.D.: Modeling Pavement Performance Based on LTPP Database for Flexible Pavements. Teknik Dergi, Dergi Park Academik, vol. 31(4) (2019). https://doi.org/10.18400/tekderg.476606, Will be published in year 2020

Shafabakhsh, G., JafariAni, O., Talebsafa, M.: Artificial neural network modelling (ANN) for predicting rutting performance of Nano-modified hot-mix asphalt mixtures containing steel slag aggregates. Constr. Build. Mater. **85**(PP), 136–143 (2015). https://doi.org/10.1016/j.conbuildmat.2015.03.060

Shahin, M.Y., Kohn, S.D.: Pavement Maintenance Management for Roads and Parking Lots. United States Army Corps of Engineers, Technical report, M-294 (1981)

Sollazzo, G., Fwa, T.F., Bosurgi, G.: An ANN model to correlate roughness and structural performance in asphalt pavements. Constr. Build. Mater. **134**(PP), 684–693 (2017). https://doi.org/10.1016/j.conbuildmat.2016.12.186

Thube, D.T.: Artificial neural network (ANN) based pavement deterioration models for low volume roads in India. Int. J. Pavement Res. Technol. **5**(2), 115–120 (2012). https://doi.org/10.6135/ijprt.org.tw/2012.5(2).115

Vepa, T.S., George, K.P., Raja Shekharan, A.: Prediction of pavement remaining life. J. Transp. Res. Board **1524**, 137–144 (1996). https://doi.org/10.1177/0361198196152400116

Zimmerman, K.A., Testa, D.M.: An Evaluation of Idaho Transportation Department Needs for Maintenance Management and Pavement Management Software Tools (2008)

Author Index

Printed in the United States
By Bookmasters